Lindsay Kelley is a practising artist and Lecturer at UNSW Australia Art & Design.

'*Bioart Kitchen* plays with the industrial food system – taking familiar products off the shelf and making them strange. Chicken soup, Coke, peanut butter, canned food, and corn syrup will never taste the same. Kelley's collection of recipes brings feminist sensibilities to Home Economics – showing how the kitchen has long been a space of subversion, performance, and innovation.'

Eben Kirksey, Australian Research Council Fellow, University of New South Wales, and author of *Emergent Ecologies*

'This fascinating tome mixes appliance lore, technological food scares, feminist fists raised in protest, artists' pot lucks and the Neiman Marcus cafeteria into its eclectic "menu!" Study it, learn from it. If you have not already juiced your breakfast, try one of Kelley's recipes. This important read adds to a growing shelf of books that show how earlier feminist art set the stage for younger artists today engaged with social justice and food. Weaving in personal experiences, from her earliest memories of eating bologna processed into perfect circles, to the fascinating epilogue that describes her grandfather's enteral tube feeding, *Bioart Kitchen* brings intimate, alimentary, feminist and technological stands together with insight and creativity.'

Linda Mary Montano, performance artist

'*Bioart Kitchen* knots together research and display practices with the threads of art, biology, technology, and activism. I am hungry for this nourishment, and Kelley is a superb cook. The arts of eating are at stake in this book in many senses, and I stayed gladly for the full menu.'

Donna Haraway, Distinguished Professor Emerita, University of California at Santa Cruz

'Elucidated by means of a rich array of contemporary art works, this book makes new sense of the interface between what it is to cook and what happens when artists blur the contact zones of the kitchen and the gallery. With an engaging lightness of touch, Kelley speaks to the complexity and the urgency of what matters in the bioart kitchen. Read it and eat!'

Lynn Turner, Senior Lecturer in Visual Cultures, Goldsmiths, University of London

Lindsay Kelley

BIOART KITCHEN

Art, Feminism and Technoscience

BLOOMSBURY VISUAL ARTS
LONDON • NEW YORK • OXFORD • NEW DELHI • SYDNEY

BLOOMSBURY VISUAL ARTS
Bloomsbury Publishing Plc
50 Bedford Square, London, WC1B 3DP, UK
1385 Broadway, New York, NY 10018, USA
29 Earlsfort Terrace, Dublin 2, Ireland

BLOOMSBURY, BLOOMSBURY VISUAL ARTS and the Diana logo
are trademarks of Bloomsbury Publishing Plc

First published by I.B. Tauris in Great Britain 2016
This paperback edition first published 2022

Cover design: Liron Gilenberg, Ironic Italics
Cover image: *Hygge House*, 2013. A temporary installation by Plain Projects, Pike Projects and URBANINK on the Red River Mutual Trail in Winnipeg, Canada.
Image courtesy Rob Swystun, Flickr.com

A catalogue record for this book is available from the British Library.
A full CIP record is available from the Library of Congress

ISBN: HB: 978-1-7845-3413-4
PB: 978-1-3502-7094-7
ePDF: 978-1-7867-3000-8
eBook: 978-1-7867-2000-9

Typeset by Newgen
Printed and bound in Great Britain

To find out more about our authors and books visit
www.bloomsbury.com and sign up for our newsletters.

For my grandmothers, Jean Edwards and Ellen Kelley.

Contents

List of Illustrations

Acknowledgments

Many friends, colleagues and mentors have made invaluable contributions to my thinking since this work first began as a dissertation at the University of California Santa Cruz in the History of Consciousness Department. Thank you to my PhD supervisors, Donna Haraway, Warren Sack, Margaret Morse and Gary Lease. Thank you also to my Digital Art & New Media MFA supervisors, Elliot Anderson, Elizabeth Stephens, and Amy Balkin. My colleagues at Histcon have been fantastically supportive. I'd especially like to thank Eva Hayward for being the best co-author, co-organizer, collaborator and friend, Eben Kirksey for his tireless work on *The Multispecies Salon* and Katie King for nurturing an intergenerational transdisciplinary network of friendship. Thank you Amy Remensnyder and Chip Lee for teaching me about writing, struggle, and struggling to write.

I could not have completed this manuscript without the support I've received from UNSW Australia Art & Design. My position here gives me the ground from which to invest time and energy in my research. My colleagues and students across Art & Design and the Environmental Humanities are generous, rigorous and kind. Special thanks to the 2015 BFA and SOMA Honours cohorts.

Several of these chapters were first developed as conference papers. Thank you to the organizers of the Curious Lives of Documents Symposium and Midnight University (University of California Davis, 2003), Meet Animal Meat (Uppsala University, 2007), Invisible Ingredient at Rock! Paper! Scissors! (2009), Natureculture: Entangled Relations of Multiplicity at the Society for Cultural Anthroplogy's 2010 meeting, The Multispecies Picnic, Making, Meaning & Context: A Radical Reconsideration of Art's Work (Goddard College, 2011), the Santa Cruz Museum of Art and History's Third Friday series (2011), Thinking Through the Environment: Unsettling the Humanities (UNSW Environmental Humanities, 2013) and Feral Experimental (UNSW Art & Design galleries, 2014). I would like to especially thank Bob Davidson for inviting me to present *Dysphagiac* at the

University of Toronto in 2013. The time I spent in Toronto with Bob and chef Joshna Maharaj was invaluable for developing the project.

Two of these chapters have been published previously. 'Carnal Light' was written with Eva Hayward and was first published in *parallax* 66, 'bon appétit', a special issue co-edited with the marvellous Lynn Turner in 2013. The essay is reprinted here with the permission of both Eva Hayward and Taylor & Francis, and the full issue is available at the publisher's website, www.tandfonline.com. 'Plumpiñon' was first published in *The Multispecies Salon*, edited by Eben Kirksey (2014). The chapter is republished here by permission of Duke University Press. 'From Sanitation to Bioremediation' was inspired by feedback I received on a different essay written for the edited collection *What's Cooking? Food, Art and Counterculture* (forthcoming from University of Arkansas Press). Editors Silvia Bottinelli and Margherita D'Ayala Valva's thoughtful suggestions for revisions to that essay inspired this chapter.

My editorial team, including Anna Coatman, Baillie Card, Lisa Goodrum, Pat Fitzgerald and Elena Knox, have been wonderful throughout this process. I am grateful for the time and energy given to this manuscript and individual chapters by the anonymous reviewers from I.B.Tauris and the reviewers and editors from the journals and books that have previously published my work.

I am most grateful to the National Institute for Experimental Art at UNSW Australia Art & Design for a generous publication grant, which has assisted me with all aspects of finalizing this manuscript. I would also like to thank the Tissue Culture and Art Project, Christine Chin, Maya Weinstein, the Center for Genomic Gastronomy, Sally Morris, Next Nature and the Critical Art Ensemble for their time and their kind permission to reproduce the figures in this book.

My family has been incredibly understanding and accommodating as I first struggled to perfect hundreds of unsuccessful job applications, then left academic work with no small amount of angst before finally returning to the university only to ask my partner to relocate halfway around the world. Thank you, Sudhakar Chandrasekharan, Kathleen Kelley, Dan Kelley, Ellen Gordon and Virginia Anthony, for your love and support. Thank you also to my cousin Rebecca Kelley for authoring our family's first nonfiction and fiction books. Thank you for making our family a family of writers.

Introduction: What is Food?

Light golden cake

Grease and flour 2 9" layer pans or 13 × 9" oblong pan
Sift together 2 ¼ cups sifted SOFTASILK
1 ½ cups sugar
3 tsp. baking powder
1 tsp. salt
Add ½ cup soft shortening
Pour in a little over half of 1 cup milk, 1 ½ tsp. flavoring
Beat 2 min.
Add remaining milk and 2 eggs (⅓ to ½ cup)
Pour into prepared pans. Bake until cake tests done. Cool. Finish with filling or frosting as desired. The oblong or square cake is ideal for the easy Broiled Icing, Choc-O-Nut Topping, etc.
TEMPERATURE: 350 (mod. oven)

TIME: Bake layers 25 to 30 min., square or oblong 35–40 min.[1]

~

What is food? What are the limits of food? Jacques Derrida offers a compelling explanation for food's inherent relevance: 'one never eats entirely on one's own: this constitutes the rule underlying the statement, "one must eat well"'.[2] To eat well is to consider food in all its complexity. Never being 'on one's own' means never being free of webs of interdependence with other people, animals, technologies, industries and economies. Artists working with wet biology and biotechnology evoke these networks by revealing the webs of 'excessive responsibility' that wrap around even the simplest meal.[3] The political commitments of food art foreground the changes and potentials within our food supply that preoccupy critical artists engaged with biotechnologies as central methodologies. Many of these artists intend their projects to further research on agriculture, eating practices and social justice.

This book considers bioart research that takes food and eating as central concerns, allowing our sense of taste to situate visual culture, embodiment and transspecies encounter. Bioart broadens the scope of performance and installation art to include 'wet' media like living plant and animal tissues, human and animal bodies and ecological systems.[4] This book joins an expanding network of contemporary theorists, artists, curators and art historians asking urgent questions about art and biology.[5] In both the popular and academic literature, food maps the terrain of interspecies interaction, the sustainability and health of our environment and the changing technologies we bring to and consume at the table.[6] In addition to books and exhibitions, recent international conferences and journals indicate a vibrant interest in food and eating on the part of scholars.[7] In the following chapters, I show how bioart emerges from feminist performance practice, food art, domestic computing and Home Economics. Reading bioart as one result of these many entangled cultural histories allows for richer generative connections than are available when considering bioart as the latest manifestation of new media and digital art. These links rebound to broaden our understanding of new media and digital art. In this way, bioart gains hermeneutic force as a dynamic framework for translation and interpretation between disciplines and materials.

My interest in bioart arises from an interest in the kitchen as a site of knowledge production. The food we eat and its preparation define who we are culturally, socially, economically and politically. As food technologies become more sophisticated, food's capacity to deceive and manipulate increases. For example, my mother continued feeding me processed meats for years after I decided to become a precocious vegetarian at age four. Thanks to their simple shapes, I didn't catch on to fish stick rectangles and bologna circles until Mom slipped with a command to 'eat your meat'. All foods, perhaps especially meat, have the capacity to dissemble, through processing, recombination and clever engineering. I became interested in food's disguises, the ways in which foods morph from one form to another and the limits or fringes of food. As an artist, I have taken this concept of 'fringe food' as a platform for investigating alternative modes of ingestion, looking at the nutriments produced in response to starvation, in the contexts of medical intervention into dysphagia (the inability to swallow), humanitarian aid work in Africa and rationales for geophagia (eating earth or mud). My kitchen and my studio are often the same space, and I became interested in other artists whose practices demand similar work environments. I found such artists in the bioart world, where projects require hybrid spaces that are kitchens, labs and gardens all at the same time.

During my research, my mother's and grandmothers' kitchens became research sites. How are my mother, my grandmothers and I beholden to industrial food systems? Observing our different ways of cooking and eating, I realised that we are inadvertent apprentices to home economists at work in corporate test kitchens. Our knowledge has evolved alongside new developments in test kitchens. Consider baking a cake. For my mother, Kathleen Kelley, a cake mix is a convenience and a guilty pleasure. Even if she sacrifices perceived quality by using it, the mix offers her needed time. For her mother, a cake baked from a mix is exotic and exciting, while a cake baked from scratch is backward and even embarrassing. My grandmother, Jean Marie Edwards, tended to forget things. Each new appliance, each advance in packaged food, gave her permission to let go of something old and outmoded. A SuperMoist Rainbow Chip cake was nowhere to be found during her Depression-era childhood. The Betty Crocker box will inevitably make a better, newer, happier memory, divorced in its process and chemical composition from anything she might have tasted in the 1920s and 1930s. For my paternal grandmother, the endurance challenge of making massive amounts of food by hand connected her to her 15 brothers and sisters and to her mother, who baked bread for the family every week, enforcing harsh punishments for anyone who pinched the dough. Ellen Kelley never used a mix, and was proud to have the skills to reject Betty Crocker's offerings. But in the last years of her life, at her Heritage Mountain View assisted living apartment, my grandmother prided herself on her sparkling clean oven and empty kitchenette cupboards. She never baked again. She had done enough.

This peek into Denver's bungalows and nursing facilities hints at the different lessons three generations of women in the United States have learned from corporate mascots like Betty Crocker and passionate educators like Julia Child. For women of my generation, internalising the domestic science taught to our mothers and grandmothers has resulted in the emergent do-it-yourself (DIY) practices of twenty-first-century makers. Subversive kits and recipes undo Betty Crocker's capitalist grip on culinary pleasure in the process. Many contemporary DIY producers hold anti-capitalist rhetoric and boutique exclusivity in productive tension. Usually working at home, they are more likely to fund themselves from retail sales and popular press publications than from university grants. This homemade, handcrafted philosophy contrasts with a dominant food culture where products imagined by science fiction writers are now available on supermarket shelves. Artists often merge DIY with the fabulations of science fiction and other imaginings of the future to critique present-day realities of engineered food and factory farming. Visualising transspecies embodiment and the invisible labour of animals become increasingly important tasks for art in a world where fictions are quickly becoming real. Returning food production to small-batch, human scale manufacturing combats the wholesale acceptance of dystopian food futures.

Keeping simple meals and domestic kitchens in mind, this book takes the form of a cookbook, each recipe accompanied by analysis and documentation to support critical readings of bioart and its wider discursive network. The book has been designed to be read as one might read a cookbook – to be absorbed according to interest, the particular needs of one's table, and the ingredients at hand. The recipes are nevertheless ordered, moving chronologically from early histories of public amateurism to art projects that are still evolving in the present and near future. *Bioart Kitchen: Art, Feminism and Technoscience* focuses on three themes. First, I trace a history of bioart in archives of domestic labour, Home Economics and feminist art; areas usually overlooked in genealogies that position bioart as an extension of new media and digital art. Second, I analyse how 'art labs' are vital spaces for food art, providing locations where traditional 'lab studies' might merge with recent installation art practices. Third, I consider food as physical and semiotic material, positioning food and eating as central concerns for activists and artists alike.

I start with a breakfast of oatmeal, analysing how the body of the home economist contains and complicates the collapse of the kitchen into the laboratory. Chapter 1, 'Subject P: Embodying Home Economics', argues that frameworks of public amateurism at work in bioart engagements with food and eating were first activated by women scientists who pursued Home Economics at the university level. For the women enrolled in these programmes, a university degree in Home Economics presented many paradoxes. The discipline created opportunities for women that did not exist before: university-level study of Home Economics resulted in careers in test kitchens, high schools and even stints as a corporate mascot like Betty Crocker or Joan Allen. In an era when women could not pursue the chemistry or physics degrees they would have otherwise enrolled in, university-level Home Economics pushed serious science research underground by forcing women scientists to pursue topics that conformed to a discipline that produced generalists, not specialists.

Chapter 2, 'Chicken Heart Soup', navigates early tissue culture work in both laboratories and speculative fiction to expose new figurations of the animal body in pieces. A fan-created recipe for Margaret Atwood's 'Chickie-Nobs' opens the chapter, and the chicken becomes our guide to her own transformation from a barnyard animal to a laboratory specimen to a monster dwelling in the cages of a speculative future/present. Because tissue culture is a technique rather than a discipline, early practitioners like Alexis Carrel and Honor Bridget Fell integrated the process of culturing tissue into more specific research questions. This chapter gathers together these and other histories of tissue culture processes and the speculative futures of Julian Huxley, Winston Churchill and Margaret Atwood, whose fictional worlds propose a myriad of ways that tissue culture might be integrated into culinary and familial daily life. Tissue culture joins performance, speculative fiction and art in its emphasis on process over product.

Chapter 3, 'Domestic Computing', indexes the changing roles of women and machines in and outside of the home. Featuring a recipe from Helen Corbitt for Neiman Marcus, the culinary celebrity attached to Honeywell's Kitchen Computer (the first computer marketed to private consumers), the chapter follows women and machines across industry and domestic contexts. Home Economics' devotion to Taylorist efficiency models marked domestic labour as a problem to be solved for the engineers who created early domestic computers like the ECHO IV and the Kitchen Computer. Uneasily regarded by homemakers as physical and emotional replacements for their work and bodies, domestic computers ranged from empowering devices co-designed with their household users to elaborate jokes that left women out of the equation entirely. Looking at how both Home Economics and domestic computing have shaped our food and our kitchens reveals an investment in systemising women's labour in the home. These machines were unique prototypes in the 1960s, but today's 'smart' appliances and online recipe archives have integrated most of their features. Controlling lighting, heat, even windows and doors, from a centralised computer panel no longer requires the ECHO IV's enormous processor lurking in the basement. The recipes and menus painstakingly encoded and decoded from the Kitchen Computer may be accessed from any laptop with a quick internet search. The normalisation of these technologies parallels the internalisation of lessons from Home Economics; using the new foods and methods promoted by Home Economics has become as intuitive as searching for a recipe online.

Chapter 4, '*Semiotics of the Kitchen*: Feminist Food Art', finds that feminist politics have undone and troubled the efficient, Taylorist kitchens of the past. Opening with Martha Rosler's iconic list of materials and implements, the chapter locates a performance politics of food and eating in feminist art of the 1970s. I argue that canonical feminist works were animated by understandings of the body that predict contemporary interest in 'assemblages' of objects and actors: domestic architectures and tools both extend and decentre the body.[8] Marking a break from both domestic computing and Home Economics studies, feminist art overturned many assumptions from both of these fields. Although recipes were rare forms in feminist art, food and eating were central concerns. I focus on Martha Rosler's *Semiotics of the Kitchen*, Judy Chicago's *The Dinner Party* and related projects by Suzanne Lacy, as well as the collectively produced *Womanhouse* and Linda Montano's 'Food & Art' special feature in the journal *High Performance*. Feminist art responds to both Home Economics and domestic technology, insisting that women are capable of exercising their creativity outside of the kitchen while retaining an investment in the meal format. Women's art produced during American second-wave feminism of the 1970s elaborated on existing conceptions of how food, recipes and performance might combine to produce a critical art practice of political consequence. This work departed from Home Economics and domestic computing by focusing less on standardisation and efficiency, and more on how food and dining interface with political mobilisation. Feminist artists transformed eating, housework and sex from private aspects of our daily lives to public catalysts for political change.

Chapter 5, 'DIY Coke', introduces my art practice alongside the work of other artists and activists engaged with industrial intervention in the form of kits and renegade corporate recipes circulating online. Examples of such forms include Martha Rosler's *Romances of the Meal* and Maya Weinstein's *DIY High Fructose Corn Syrup*. Rosler compiles recipes for industrial baking with an emphasis on Coca-Cola and coffee. Extending from her early video *Semiotics of the Kitchen*, Rosler reveals industrial kitchens that are just as hostile and alienating as her minimalist video set. The chapter begins with a recipe purporting to reveal the secret formula for Coca-Cola, and follows that recipe as I prepare ten litres of bootlegged Coke during a small *DO-IT* exhibition at the University of California Davis. Weinstein has created the first in a series of kits related to processed foods, *DIY High Fructose Corn Syrup*, to bring industrial food products like HFCS into home kitchens. All of these projects take advantage of the spaces opened up by the feminist art movement while simultaneously working to politicise locations that are only beginning to be critically interrogated from a feminist perspective, like industrial kitchens, factories and corporate research and development labs.

Chapter 6, 'Meat Culture', begins with a recipe for the care and handling of fetal bovine and fetal calf serums (FBS/FCS), the most important ingredients in contemporary tissue culture protocols. I consider contemporary *in vitro* meat research and the 'victimless' utopias of the Tissue Culture and Art (TC&A) Project. *In vitro* meat has long been a staple of speculative fiction – visions of the future always include strange new meat delivery systems, from cannibalistic processed Soylent Green to monstrous chickens featuring dozens of breasts stacked around an immobile oesophagus. Just as recipe art shows us how transgenic futures are closer than we might think, the TC&A Project's *Disembodied Cuisine* reveals the present-day manifestations of speculative fiction, invoking the ways in which we are already manipulating livestock to extreme ends. Monstrous immobile chickens, for example, flop around broiler sheds all over the world, engineered to privilege breast meat to such a degree that they can no longer walk more than a few steps, if they can stand at all. Are broiler chickens kin to the TC&A Project's 'semi-living' creations? Would we prefer to raise *in vitro* chicken meat while donor chickens run free? In all of these meat production scenarios, we are forced to consider how human and non-human animals encounter one other in the lab, the slaughterhouse, and the dining room. With this chapter, the book turns to bioart as such, no longer identifying precursors and parallels. The works analysed in the second half of the book were emergent, exciting and new when I first started writing about them in 2003. In the ten years since those first explorations, these works have become representative of this now established genre of art practice. Rather than reinscribe a canon of bioart, I aim to revisit these works in light of their resonance with diverse histories of feminist food art and Home Economics, suggesting that their status as representative works may owe something to their engagement with this interdisciplinary prehistory.

Chapter 7, 'Public Amateurism', follows the Critical Art Ensemble's (CAE) efforts to open up the processes of biotechnology with a series of public interventions. In their collaborations with scientists and the public, Critical Art Ensemble members work as 'public amateurs', showing their learning process both in their written work and the public installations of their projects. These installations examine the stakes of genetically modified (GM) food, interrogating the particular politics of this form of biology by way of participatory projects designed to illuminate how we are implicated in industrial agriculture. *Molecular Invasion* 'reverse engineers' genetically modified cash crops patented by Monsanto, proposing 'fuzzy biological sabotage' with the help of an enzyme inhibitor developed by CAE and 'rogue biologist' collaborators. *Free Range Grain* offers visitors the chance to have their foods tested for GM markers, responding to debates in the European Union about food labelling and the fantasy of a GM-free Europe. CAE advocates a range of tactical interventions into corporate and regulatory spaces, flirting with the rhetoric of sabotage and terrorism. Interpreting their DNA extraction protocol from *Free Range Grain* as a recipe for biotech kitchens, this chapter finds that of all the projects in the book, CAE's work most directly inherits the legacy of home economists, particularly the work of early practitioners engaged in public demonstrations. New workspaces helped create public amateurism. I call these hybrid laboratory-kitchen-garden-studio spaces 'art labs'.[9] In CAE's practice, art labs may emphasise one or another element, sometimes remaking an existing research laboratory into an art space by virtue of the materials and activities taking place there, at other times resembling the home laboratories celebrated by nineteenth-century amateur naturalists. The capacity to perform research, create new media and display biological and media research *all at once* distinguishes the art lab from previous definitions of studio-laboratory hybrids.[10] These spaces instigate ruptures, especially around labour and gender, as art labs often collapse feminine domestic interiors like kitchens and dining rooms with the more traditionally masculine sphere of the research laboratory.[11] Critical Art Ensemble's collaborative projects collapse the kitchen and laboratory bench to produce new subject positions and new relationships with the history of laboratory culture.

Chapter 8, 'Cookbook', investigates the cookbook format as a vehicle for political critique. Turning first to art and activist critiques of genetically modified foods, invisible technologies and ingredients are made visible with recipes like Christine Chin's 'Garrulous Gazpacho'. Chin's *Genetically Modified Foods Cook Book* calls for fantastic ingredients like finger carrots and mouthy tomatoes, all of which are vegetable human hybrids that meld human sense organs with vegetable forms. Chin's transgenic foods play with cannibalism, hygiene and bodily integrity, asking cooks to care for the uncanny vegetables as we care for our own bodies. Next Nature's *The In Vitro Meat Cookbook* presents both speculative and realised visions of cultured meat, joining Chin in an effort to visualise food cultures of the future. Speculative and historical recipes reflect changing food industry standards, diet fads, commercial products and technological innovations. Some recipes assume more than others: early recipes remind the reader of ingredients without spelling out precise amounts or how a dish is put together, while today's recipes detail both ingredients and protocol.[12] While men still dominate the kitchens of elite restaurants in many parts of the world, women are our favourite recipe keepers (postwar icons of the recipe form include both corporate mascots like Betty Crocker and Aunt Jemima and real cooks and cookbook authors like Julia Child and Helen Corbitt).

Recipes circulate in many different contexts, and have been deployed for diverse purposes. Beginning in the 1950s, FLUXUS artists adopted the recipe format as a way to document their work and to make it reproducible by other people. Recipes are documentation, security against forgetting, traces of culinary success and failure: recipes remain after the meal is over. Drawing from the recipe protocols and test kitchens developed in Home Economics programmes and from feminist art's tradition of politicising the personal, recipe artists bring challenging, unfamiliar ingredients into the most quotidian, comfortable books on our shelves: cookbooks.

Written with Eva Hayward, Chapter 9, 'Carnal Light', takes another canonical bioart project as its starting point. Turning away from the celebrity circulations of Eduardo Kac's *GFP Bunny* and towards the material semiotic tracings of the jellyfish protein that prompts her to glow, this chapter reads Alba as 'somalumenal' and 'tranimal', or trans-animal, an organism animated by trans-formations of life. The Critical Art Ensemble's recipe for cooking Alba, 'Ragout Alba a la Provençale', opens up the troubling space between pets and meat. CAE evokes Betty Crocker's role as a proponent of new, industrial ingredients: putting green fluorescent protein (GFP) on the dinner table engages the recipe form while offering a critique of transspecies genetic engineering and industrial food production. We distinguish the politics of visibility from the technologies of visuality, asking how the bodies of GFP-donor jellyfish are visualising but invisibilised forces, even as GFP-host animals steal headlines. GFP hosts and donors reveal the carnality and radiance of naturecultures by welcoming us to a messy, multidisciplinary table where we are eating light.

Chapter 10, 'From Sanitation to Bioremediation', continues to follow material networks of connection between organisms and ecosystems, turning to projects that promote unlikely sensory paths into ecological crisis. These include Helen Mayer Harrison and Newton Harrison's *Survival Piece No. 2: Notations on the Ecosystems of the Western Salt Works (with the Inclusion of Brine Shrimp)*, Mel Chin's *Revival Field*, the molecular gastronomy of Natalie Jeremijenko's *Cross(X)Species Adventure Club* and the Center for Genomic Gastronomy's *Smog Tasting*. I argue that contemporary engagements with bioremediation, mapping and environmental intervention inherit their emphases on visualisation and public accountability from early public health advocates practising 'sanitation'. Together, these works push the technoscience of food beyond the immediate aesthetic impressions of processes ranging from phytoremediation to molecular gastronomy. Experimental outcomes from these works result from cultivating, harvesting and ingesting specific technologies.

I conclude the book with two essays grounded in my art practice, which centres on critical recipe development, fringe foods and alternative modes of ingestion. Bioart projects about food have always pushed the boundaries of what it might mean to eat and cook within the expanded field of biomedia. The two projects that close the book, *Plumpiñon* and *Dysphagiac*, respond to different forms of scarcity and starvation. *Plumpiñon* puts the piñon seed or pine nut in dialogue with the corporate humanitarian aid food Plumpy'nut, dropping into the post-colonial networks that animate starvation narratives across time, continents and empires. *Dysphagiac* invites the person who cannot swallow back into the kitchen by designing a speculative kitchen appliance that quotes laminar flow cabinetry principles. Performing the mechanics of chewing and swallowing outside the body opens up the mechanical and microbial systems that make eating possible for the swallower and non-swallower alike.

To understand how laboratories, kitchens and dining rooms intersect in the bioart practices of the last 20 years, these chapters have turned to histories that initially seem far removed from familiar and valuable readings of bioart as an outgrowth of digital media and the technologies of genetic engineering.[13] While bioart does resonate with and relate to many digital art, cyber art and new media art practices, the cultural concerns of bioart have a long, broad history that includes food, eating, labour and the home. Home Economics, second-wave feminist art and recipe art expand and enrich the bioart landscape, illuminating connections between scientific research and development, domestic interiors and amateur labour. Contemporary art practices that critically engage with biotechnologies and food share workspaces and research architectures with domestic science, in its institutionally-sanctioned and amateur manifestations. Because many food art pioneers are feminist artists, I read bioart as a practice informed by feminist and ecofeminist methodologies, even if bioart projects do not always have explicitly feminist aims. Kitchens are often conceptualised as labs. As more and more of our food supply derives from corporate research and development, labs and kitchens necessarily overlap. Concerns about social justice, food justice and the work of women and non-human animals animate both my own analysis and the projects undertaken by food artists past and present.

In exploring alternative contexts for bioart, I join theorists and practitioners who are frustrated by the narrow reach of art historical narratives. In her doctoral thesis, 'Growing Semi-Living Art', tissue culture artist Ionat Zurr outlines the dominant story told about bioart:

> Many scholars draw a direct line from genetic art [...] to BioArt [...] In order to rationalise this leap from computer-generated art to art that involves the manipulation of biological life, the proponents of such a narrative take the view that biological life is all about the code; that the artists and the work involved with biological art deal with the 'code' of life. One can speculate that the combination of genohype and the need for cohesive narrative leads to ignoring the complexity of the different levels of engagement with life. This proposition leads to an assumption of a linear, controlled, and progressive history of BioArt that seems to be the line of choice of most art historians, curators and theorists, who find it hard to cope with the multiplicity of sources, concerns, motivations, backgrounds and references of BioArt.[14]

Zurr finds digital art origin stories inadequate, given the inability of such narratives to represent the diverse backgrounds of bioartists and the wide range of content they pursue in their work. This book unsettles the dominant narratives Zurr laments by connecting bioart projects with archives that predate digital and new media art. By turning to recipes, histories of food art and feminist critiques of food and eating, I contribute to productive multiplicities that complicate intuitive interpretations of biomateriality in art.

Throughout the book, I focus on the multisensory reception of food and art, the shared work environments of artists and scientists and the hijacking of corporate design as an aesthetic of resistance. I am interested in physical and discursive encounters between artists, scientists and activists, and in the changing definitions of the 'amateur' and the 'embed', as the terms are used across art, science and even military contexts. The following chapters find that our daily meals are deeply influenced by Home

Economics and the increasing prevalence of genetically modified food and organisms. The changing modalities of food art now include street-level interventions and installations where watching trees grow and cells multiply punctuate and complicate the 'live' in live art. These interspecies performances enact sensory experiences that work not only visually but also in the realms of taste and touch. What does it mean to eat when technologies are being eaten? How do technologies taste? Hidden, involuntary responses like swallowing, gagging and the rumblings of digestion are also at play in these projects, compelling us to read the work in haptic and proprioceptive registers. We do not so much see as digest the recipes collected in this book.

1

Subject P: Embodying Home Economics

The oatmeal was [...] mixed [...] with the 95 gm. of cornstarch and added together with 40 gm. of butter fat and 5 gm. of salt to 440 gm. of hot water. This was cooked directly over the flame until a thick mush was obtained which was spread on pans in sheets a to ¼ to ⅜ of an inch thick and put in a Freas electric oven for a period of 2 hours and, when done, was quite dry and crisp throughout but not brown.[1]

~

Excerpted from a technical paper in *The Journal of Biological Chemistry*, these instructions detail the preparation of oatcakes for a study on maximising caloric intake from oat protein. Tasked with consuming this strict diet was Subject P, also known as Velma Phillips, a young Home Economics graduate student who took her kitchen as a laboratory where she could experiment on her own body. Joining test kitchen cooks and dieticians, Subject P stands in for a fictional underprivileged subject as she transforms her flesh with exacting preparations of oatmeal.

Promoting efficiency and universal standards for everything from individual kitchens to the mass production of food in factories, Home Economics set the stage for many of the domestic technologies we use on a daily basis. This chapter reads Home Economics as an interdisciplinary field, paying special attention to the contradictions faced by women undertaking Home Economics research in a university setting. University Home Economics departments housed serious work in biology, chemistry and physics, but these projects were rarely acknowledged as such. Although Home Economics was an interdisciplinary and expansive field, women working on even the most technical projects had to adapt their research content and methodology to the domestic context. While university women contained their research in the home, homemakers began to see their work as research. A form of public amateurism develops from Home Economics that will later be taken up by feminists and bioart practitioners engaged with food and eating. Home Economics sets the table for a long and provocative meal, serving up feminism, technoscience and domestic technology to eventually arrive at what Robert Mitchell calls the 'problematic of biotechnology'.[2] Before engaging this and other problematics, let us look first at the labour and bodies of home economists, among them some of the first women in science.

Among university Home Economics programmes, Cornell's Home Economics Department offers one of the first, biggest and best-documented examples of Home Economics in the United States. What began as a reading course became a degree-granting co-educational department in 1907, becoming the New York State College of Human Ecology in 1969. Even though the Morrill Act of 1862 provided for co-educational land-grant colleges, women were not earning degrees from these institutions until much later. Because Cornell's Home Economics programme began as a reading course for farmers' wives, degrees were not seen as logical or desirable outcomes for participants – the non-credit reading course continued until 1921.[3] Decades later, after establishing Home Economics as a department, professors Martha Van Rensselaer and Flora Rose were the first women to be granted full professorships at Cornell. The American Home Economics Association was founded only a year after Cornell's programme began, in 1908. Its first president, Ellen Swallow Richards, was the first woman to graduate from the Massachusetts Institute of Technology, where she studied and later taught chemistry.

Richards promoted scientific engagement with domestic labour in books like *The Chemistry of Cooking and Cleaning*,[4] and her Rumford Kitchen, a model kitchen installed at the 1883 Chicago World's Fair.[5] Had MIT been willing to grant a woman a graduate degree, she would have earned her doctoral degree in chemistry. Even without her PhD, she taught at MIT for many years (she was unable to find a job elsewhere), where she developed a laboratory space for women designed to accommodate diverse engagements with chemistry.[6] While working in MIT's chemistry laboratory before she had access to her own space, she wrote that she was 'segregated in a special corner, "very much as a dangerous animal might have been"'.[7] Perhaps as a means of assuring her co-workers that she was not such a dangerous animal, Richards and her students at the Women's Laboratory chose gender-appropriate subjects: the home, domestic work and 'sanitation', a precursor of public health, which could be interpreted to include the urban systems Richards worked with extensively (especially the municipal water supply) and broader societal concerns about morality and family life.

As Home Economics developed within land-grant college curriculums and spread to other university settings, the undergraduate degree developed areas of concentration, including clothing and textiles (pattern making and home sewing grew in popularity alongside the Home Economics project), child care, hospitality, household equipment and technology development and food science. Mary Drake McFeely characterises Home Economics departments as institutional manifestations of Richards' isolation in MIT's chemistry lab: 'Wary of female professors in traditional science departments, university administrators used Home Economics as a convenient place to segregate women scientists and keep them from competing with men.'[8] This separation had the effect of focusing home economists on educational rather than industry opportunities. Although a range of jobs existed within Home Economics, teaching and homemaking dominated the field, with a large number of Home Economics graduates holding high school or university level teaching positions at one point in their careers.[9] Many students of Home Economics chose to 'enter marriage' and embark on homemaking as the unpaid career they had been preparing for with courses devoted to cooking and laundry. Although Home Economics departments were either phased out or renamed by the 1970s, the 1950s and 1960s produced innovative if at times startling Home Economics experiments, including Cornell's practice apartments, where students honed their homemaking skills with babies borrowed from local welfare agencies.[10]

As a discipline, Home Economics gave significant numbers of women opportunities to conduct scientific research in a university setting but continuing into the twentieth century such research was constrained by topics appropriate to domestic science, like cleaning chemistry or household efficiency testing.[11] The discipline's commitment to food and nutrition allowed for a level of scientific seriousness and rigour unavailable to home economists focused on, for example, sewing and patternmaking. Nancy Berlage identifies the emphasis on nutrition as a pivotal professionalising move for the discipline and for women scientists: 'By moving into nutrition, home economists made a statement that science was indeed a proper realm for women [...] they transformed "cookery" and related experimentation in food preservation and conservation into nutritional science.'[12] Despite these tactical advances into professional territory, the home economist was often labelled 'amateur', but not in the progressive sense valued by today's bioartists. She was rarely able to attain professional status in her workplace because Home Economics departments produced generalists, not specialists.[13]

Despite the limits imposed on their professional accomplishments, home economists have influenced how the general public performs nearly every activity in the home. Outside of the Home Economics classroom, everyone who works in the home, from working mothers to professional housecleaners, is unwittingly apprenticed to home economists and test kitchen technicians.[14] Every time I follow the instructions on a cake mix, I am the inadvertent student of a team of researchers responsible for analysing the chemical, psychological and caloric properties of my finished cake. Later in this book, the artists I read as 'recipe artists' critically reinvent and reinterpret Home Economics pedagogy and principles, transforming our engagement with its subtle, internalised networks into explicit and conscious activities.

In addition to changing how we work, Home Economics changed our understanding of where we work when we work in the home. Household tasks can be interpreted as research, and the kitchen can be a laboratory. Who is allowed to work in the laboratory versus the kitchen? How do women convert one into the other, and how does the artist's studio extend and subvert both? Kitchens and laboratories are locations of cultural stress and struggle between genders and between conceptions of the self and conceptions of the other. Feminist art, for example, asks how feeding others and the self can be taken up as political activities. Within both university and corporate laboratory spaces, women have operated in covert ways to produce hard science disguised as Home Economics. They have taken their own bodies as scientific instruments, conducting food studies and quantitative analyses that return decades later in critical art projects like Eleanor Antin's *Carving: A Traditional Sculpture* (1972), and Martha Rosler's *Vital Statistics of a Citizen, Simply Obtained* (1977). Recipe art highlights the simultaneously expansive and constrained femininity of the form, shifting perspectives on both history and collaboration, and producing kitchens and laboratories as gendered workplaces.

Recipe art has a long and varied history which includes Home Economics as interdisciplinary science practice, domestic computing and 'kitchens of the future' and feminist interpretations of eating and food preparation. If domestic computers and Taylorist models of efficiency write women as mechanical brides, feminist art about food and eating seeks to divorce women from simple tasks and chores while simultaneously insisting that women's work is legitimate labour. Tasks and chores have full rich histories of their own, and their histories were made legible by the Home Economics project, which applied scientific research principles to what had previously been cloistered, private work. Ruth Schwartz Cowan writes a history of domestic technology in her book *More Work For Mother: The Ironies of Household Technology from the Open Hearth to the Microwave.* Cowan argues that the advent of home appliances mechanised tasks that men usually performed, making 'more work for mother' because women had to spend time and energy managing these machines, forfeiting male labour in the process.[15] Kitchen computing and mechanical recipe indexes serve similarly ironic roles. As various kitchen controls are relinquished to machines, 'smart' appliances enlist women in managerial processes that range from drudgery to creative contributions.

Books like Christine Frederick's *Selling Mrs Consumer* (1929) carved out a space for home economists in corporate America.[16] Women working in corporate jobs 'played a negotiating role between consumers and corporations', with increasing opportunities to become involved in Home Economics in a corporate context during the interwar years in the United States.[17] A surge of such opportunities characterised the late 1950s and early 1960s, decades when Home Economics flourished before its identity crisis arrived alongside second-wave feminism in the late 1960s and 1970s. Two books published in the late 1950s and early 1960s, reveal the ways in which women used Home Economics degrees to pursue careers outside the home. Velma Phillips's *Home Economics Careers for You* (1957) and Jeanne Paris's *Your Future as a Home Economist* (1964) encourage women to consider pursuing university education.[18] Both books follow Home Economics graduates into a variety of professions, including recipe development at the corporate level.

Phillips profiles Kay O'Neal, Director of Home Economics at Kroger. O'Neal assumes a collective corporate mascot identity, Jean Allen. Allen exists as four separate people who represent Kroger at public talks and cooking demonstrations. The Jean Allens also work behind the scenes 'testing and developing recipes to be printed on Kroger labels'.[19] Betty Linn works in the Mix Research and Development Division at Quaker Oats. Her laboratories produce both baking mixes and accompanying instructions.[20] Jeanne Paris profiles Joy Grawmeyer, executive dietician in the Menu and Food Standards Department of Linton's Restaurants. Grawmeyer 'dreams of test kitchens', where she works with an assistant to 'develop new recipes, products, and equipment in our research kitchen'.[21] Grawmeyer rose to her executive position quickly after graduating with her Home Economics degree in 1955, and is the only woman working within a managerial role in her company. The test kitchen is one of the most important laboratories available to home economists. Recipe developers use test kitchens to refine standards for both the chemistry of food and efficiency in its preparation. Recipe developers typically work for large corporations, but test kitchens also existed in university settings and at travelling educational demonstrations, like Richards' World's Fair Rumford Kitchen.

Recipe development and cooking often disguised biology and chemistry research topics. Among the Home Economics Masters and PhD theses filed in 1954–5 in the United States, the following titles suggest that chemistry and biology thrived within Home Economics contexts, even if practitioners tended to wear aprons rather than lab coats: 'Determination of meringue slippage and liquid drained from pies prepared in quantity' (Shirley A. Felt, Cornell, 1955), 'Lipid metabolism and cell division' (Elinor Levin, Wayne University, 1955), 'Effect of xanthophyll on the palatability, fat stability, and histological characteristics of fresh and frozen broad-breasted bronze turkeys' (Burnadine L. Lewis, Kansas State, 1955) and 'The effect of methods of preparation on the retention of ascorbic acid and dehydroascorbic acid of raw and cooked vegetables' (Lois Ann Lund, University of Minnesota, 1954).[22] These titles also reveal the importance of corporate connections for Home Economics as a discipline, showing that home economists engineered their research to be relevant to potential employers.

Although Home Economics brought many women into higher education who may not have otherwise pursued university study, the field emerged as a catchall for a wide range of ambitions, from marriage to teaching to chemistry to biology. Home Economics might be considered the first interdisciplinary degree available in the United States, as every Home Economics degree demanded the melding of diverse specialisations and practices, as well as the creation of an area of focus within a vast topography of disciplinary directions. The mechanisation of women and their work also drove the field forward, frequently putting women in the position of analysing their own bodies, movements, and choices. Velma Phillips narrates a striking example of this self-reflexivity:

> Another interesting experience I had as a graduate student was doing a metabolism study for Doctor Henry Sherman on the value of the oat protein in human nutrition. For six weeks I lived on oatmeal, with only apples and butterfat added, to find out how adequate the protein in oatmeal is for human nutrition. This experiment was financed by the Association for Improving the Conditions of the Poor [...] I worked with low-income families who had to spend half of their food money for cereal in order to get enough calories at the lowest cost. This study was published in the *Journal of Biological Chemistry* by Doctor Sherman and me – a distinct honor for a young graduate student.[23]

Although Phillips emphasises the humanitarian benefits of her oatmeal trial, this outcome of the study falls away from Sherman's article, where Phillips exists only as 'Subject P', her progression through the study noted in several tables and charts, her person mentioned only once, when she is characterised as a 'healthy young woman weighing 67 kilos'.[24] Subject P prepared her oatmeal herself, in mush and cake form, following strict protocols. Phillips' position within this study presents an extreme example of how Home Economics students conceptualised their own bodies as research data that might stand in for larger, absent groups of people (the 'poor', in this case, or housewives, nurses, teachers, in other cases). The changes the home economist might effect at the level of her habitus could eventually be propagated and spread to the bodies and behaviours of others. The home economist holds and embodies data visualisation, chemical compounds of starches and sugars and disciplined weight loss in service of the disadvantaged.

While acknowledging the difficulties of doing one's chemistry research in a test kitchen, the popularity of Home Economics degrees should not be read as a lament for the absence of women in other fields. Home Economics helped women enter university systems previously closed to them. Home economists had their own laboratories, which proposed a new model for the American kitchen: the rigorous, scientific test kitchen. Home Economics literature encourages women to regard their domestic work as scientific research:

> There are unlimited opportunities for home economists to follow [...] careful research procedures to determine the best way to carry on their daily activities. Planned experimenting to find the best way to squeeze an orange, wash a sweater, heat a baby's formula, clean a kitchen wall, or teach candy making would be research.[25]

Any kitchen can be a test kitchen, and any food preparation can be research.[26] Recipe development enacts precisely this type of scenario, making private kitchen activities public. In the 1960s, Home Economics looked outward, focusing less on the American homemaker and more on global concerns about public health, hygiene and nutrition.[27] Although this transition cannot be directly linked with any single cause, new technologies in the home and increasing access to sciences other than Home Economics may have contributed to the discipline's shift toward community rather than domestic service. This same shift promotes an early form of public amateurism. From Ellen Swallow Richards' Rumford Kitchen to the invitation to compare the private home kitchen to the corporate test kitchen, Home Economics preferred public learning to corporate secrecy.

Even so, Home Economics and feminism clashed as the latter either abandoned or was hampered by its ties to the former. Home Economics burdened the feminist movement, to the degree that when addressing the American Home Economics Association convention in 1972, Robin Morgan declared that: 'As a radical feminist, I am here addressing the enemy.'[28] Dena Attar's scathing study of Home Economics classes in British secondary schools, *Wasting Girls' Time: The History and Politics of Home Economics*, argues that 'Home Economics has a case to answer – not the obvious charge of sexist indoctrination, to which it pleaded guilty long ago, setting out to mend its ways, but a more fundamental case of causing general educational harm.'[29] This 'general educational harm' happens by limiting the subjects girls are likely to pursue, and by occupying their time with subjects that replace more advanced studies in maths and science but do not reap the same intellectual or professional rewards.

Despite its shortcomings, Home Economics complicated easy associations between women and domesticity, raising the homemaking path to the status of a legitimate career for which one trained and prepared. The labour of housework became a public concern, with household manuals like those written by Catherine Beecher in the mid-nineteenth century setting the stage for the radical assertion that women's work in the home was legitimate work.[30] Home Economics gave women a shared visual vocabulary, dictating what it meant to keep house well, cook well and nurture a happy family. Reinforced in print, film and on television, the fantasy worlds conjured by home economists were ready to be explored and exploded by feminist artists, but were also ready to be exploited by futurist (male) engineers looking to design, control and automate women's work, sometimes in elaborate, occasionally absurd ways, as illustrated by early 'smart home' appliances. The limitations of both domestic computing and Home Economics provided valuable contexts for feminism and feminist artists by professionalising crafts and chores previously discredited as private concerns and taking the first steps towards making the personal political – or at least public.

2

Chicken Heart Soup

Buffalo Chickie Nobs

What You Need!

1 cup vital wheat gluten
¾ cup whatever your favourite veggie chicken-flavoured broth is, for mixing
Some slices of onion
6 cups of whatever your favourite veggie chicken-flavoured broth is, for COOKING.
Some flour (about like […] ⅓ cup? I dunno […] enough to cover the ChickieNobs before frying)
Some vegetable oil (again I dunno […] enough to deep fry the ChickieNobs)
1 ½ cup unsalted butter
12 fluid ounces (approx one bottle) of hot sauce. Franks Red Hot is the traditional choice for traditional Buffalo Sauce […] however Chicken is the traditional choice for Chicken Wings […] so I guess you can use something different if you want to.

To Make the ChickieNobs!

Mix the wheat gluten and the ¾ cup of broth together. You can use a fork or something at first, but you will eventually have to knead it with

your hands. It will be the consistency of rubber when it's thoroughly mixed (yum!)
Break off small chunks and flatten as much as possible (about ⅓ inches thick is good). I saw a recipe on the internet that called these chunks 'cutlets' [...] which for some reason I find awkward.
Bear in mind that they will grow twice as big after they have been cooked.
Heat up the broth in a large pot, until it is boiling. Add onion. Add the flattened gluten chunks to the broth and cook for about an hour. They will expand and solidify. When they are finished remove from heat. If you do not plan to use all the seitan for ChickieNobs (it also is excellent in stir fries) than store it in the broth in your fridge.
After the seitan has cooled, remove some from broth, pat dry, and cut it into approximately chicken-wing sized bits. Dust the bits with flour, and deep fry until they are crispy on the outside.
While you are frying the ChickieNobs, melt butter in a small saucepan. Slowly add hot sauce to the melted butter, and whisk together. Remove from heat. This is buffalo sauce.
When the ChickieNobs are finished frying, toss them in buffalo sauce and serve![1]

~

Honor Bridget Fell (1900–86) and Alexis Carrel (1873–1944) were among the first scientists to systematically culture tissues outside of organisms. Working *in vitro*, or under glass, Fell and Carrel created methods and theories that would unsettle and reform conceptions of life, death and the boundaries of the body. Tissue culture is a technique and a field of knowledge rather than a discipline; both Carrel and Fell integrated the process of culturing tissue into larger research questions. Tissue culture emerges as a practice, a way of seeing and, borrowing artist and curator Phil Ross's vocabulary, a 'biotechnique'.[2] This chapter moves from past histories of tissue culture processes to current experiments with new forms of life, going on to consider the speculative futures of Julian Huxley, Winston Churchill and Margaret Atwood, where tissue culture has been integrated into culinary and familial daily life. Following another path into the kitchen laboratory, the materials in this chapter ask how food and eating change when *in vitro* technologies are being eaten.

Tissue culture was developed and advanced in terms of methodology rather than specific findings, extending Ross Harrison's early work with amphibians. Introducing her book *Culturing Life: How Cells Became Technologies*, Hannah Landecker often uses words like 'medium' and 'event' to describe the evolution of this set of practices; tissue culturists and artists share this vocabulary and their attention to process. Landecker attributes this growing emphasis on process to Alexis Carrel, 'a Franco-American surgeon who quickly switched the focus from the specific findings of Harrison's careful experiments to the method and its possibilities'.[3] By practicing and theorising tissue culture as a methodology, tissue culturists paralleled increasing attention to process in the art world. Today, the methodology's emphasis on process, performance and technique makes tissue culture especially compelling for artists, with collectives like the Tissue Culture and Art Project and Bioteknica merging the parallel histories of endurance and time-based performance art with tissue culture protocols.[4]

In her 2008 dissertation 'Growing semi-living art', Ionat Zurr addresses the uneasy position Carrel's aesthetic sensibility and politics occupy in the Tissue Culture and Art Project's work:

> While the TC&A project abhors Carrel's mysticism and belief in eugenics, and considers his aesthetic accomplishments the poor cousin of his science, his work cannot but help set the tone for aesthetic engagements with tissue culture [...] for this reason, the design of the TC&A laboratory, used in the 2003 exhibition L'Art Biotech in Nantes [...] referenced Carrel's laboratory where the first successful tissue culture experiments were performed in 1910 [...] TC&A, through its ritualization, seeks to create an intellectual and emotional situation in which the act of caring or neglecting life, even partial life, is not devoid of self reflection regarding the act and what it symbolizes.[5]

Quoting Carrel's unusual aesthetic interventions (the shadowy, dark interiors of his laboratory, his theatrical attention to presentation) allows the Tissue Culture and Art Project to focus on the aesthetics of care. This intellectual and emotional commitment to partial life engages what Honor Fell called the 'tissue culture point of view'. Fell spent most of her career at the Strangeways Laboratory in Cambridge, which she grew into a major research institution partially funded by the

Rockefeller Foundation. Susan Squier describes 'the tissue culture point of view' as 'a phrase Dr Honor Fell coined to refer to the scientific and popular mindset bred by the microprocesses of this novel technique'.[6] For the Tissue Culture and Art Project, adopting the tissue culture point of view means imagining what specific cells contribute to specific projects. Imagining the importance of tissue culture to the living organisms saved from slaughter by *in vitro* meat also entails adopting the tissue culture point of view. The 'mindset' that allows for the tissue culture point of view demands identification with tissues and culturing processes. Identification might mean forming affective bonds, acknowledging interdependence between tissues, human, and animal bodies or, in a more literal sense, extending the body by contributing one's own flesh to a tissue culture project. Scientists and artists who adopt the tissue culture point of view share the situations of the cultures they work with, both discursively and physically.

Tissue culturists care deeply about time, duration and the relationship between endurance and life. Carrel wrote that 'time is really the fourth dimension of living organisms [...] Cell colonies, or organs, are events which progressively unfold themselves. They must be studied like history.'[7] For Carrel, biological time was different, separate and, according to Landecker, most analogous to cinematic time and movement.[8] Like a classic film, always screening somewhere, always alive in the popular imagination, Carrel's most celebrated culture, the 'immortal chicken heart', lived for over 30 years, attracting enough attention that Carrel hired staff to archive press coverage.[9] It cannot have been an accident that immortality was granted to a bird we relate to primarily as food. The chicken's familiarity and edibility contribute to both its celebrity status and its influence on other thinkers invested in *in vitro* meat (Winston Churchill, for example). Long after the chicken heart's lifetime (1912–46), researchers speculated that the chicken heart was replenished with fresh cells each time it was infused.[10] Although Carrel and his contemporaries suspected the regenerative effects of 'embryo juice', the inexact term for the substance the chicken heart 'drank' each day, the persistence of the chicken heart nonetheless provided a rich philosophical terrain for Carrel.

Henri Bergson's work on time and endurance influenced Carrel's ideas about biological time. Of actions and reactions, Bergson writes, 'the greater or lesser tension of their duration [...] expresses, at bottom, their greater or lesser intensity of life'.[11] That which endures lives, and the different ways that semi-life and living things endure create a complex, interdependent network of meshing and separating trajectories. The Tissue Culture and Art Project posits the semi-living as a path into a rich, multidimensional world: 'new classes of object/being [...] will contain different degrees of life and sentience, new relationships will be formed with our environment, and with the concept of life itself'.[12] Carrel sought to extend Bergson's view of time (in Carrel's words, 'instantaneous pictures which stand out against a continuously streaming background')[13] to include physiological aspects of organs and organisms. Carrel's careful articulation of different time signatures for different organisms parallels Tissue Culture and Art Project's concern for how new forms of life and sentience will unfold. Both Carrel and the Tissue Culture and Art Project approach living matter with a commitment to what Robert Mitchell would call 'experimental vitalism'. For Mitchell, the 'vitalist mode' of bioart emphasises 'processes of life' and 'embodied exploration[s] of biological possibilities'.[14] Carrel's experiments with the biological possibilities of time grow into the Tissue Culture and Art Project's experiments with new organisms, their ways of being in the world, their tensions and environments extending established understandings of living matter and vitality.

Carrel, like his immortal chicken heart, was encased in his own nourishing environment. In *The Culture of Organs*, Carrel and Lindbergh constantly reference their laboratory décor, but never explain or justify the choice to paint the walls dark grey and outfit staff with black surgical gowns. Even the sterile sheeting in the operating room was black. This choice, which is ostensibly utilitarian, had aesthetic effects that could not have been accidental; Carrel's contemporaries found the theatrical space of his laboratory uncomfortable if not actively embarrassing.[15] In their book *Who Should Play God? The Artificial Creation of Life and what it Means for the Future of the Human Race*, Ted Howard and Jeremy Rifkin imply that Carrel intended for the black clothing to be worn 'in the presence of the immortal chicken heart'.[16] This reading seems melodramatic, even for Carrel, who wore black clothing 'in the presence of' all his tissue cultures. Carrel's own commentary about his black gowns and grey walls is brief. In the section of *The Culture of Organs* called 'Preparation of the Operating Room', he writes that:

> The operating room is the last of a suite of four rooms [...] These rooms are painted dark gray. The floor and the furniture are black. The ceiling consists of a large skylight [...] In the third room, the assistant, and later the surgeon, don their masks [...] A black, sterile gown is put on, and sterile rubber gloves [...] Double, sterile, black covers are put on each table.[17]

These remarks about the laboratory's appearance are interspersed among a larger set of instructions detailing the proper way to prepare the laboratory for surgery. Carrel laments the 'less satisfactory conditions that are ours', yet concludes that, despite the lack of filtered air, his laboratory produces 'nearly perfect results'.[18] Carrel worked for many years at the Rockefeller Institute in New York; his move to the bright, skylit top floor of the building precipitated the dark, shadowy interiors that inspired the Tissue Culture and Art Project. Carrel's objective was to absorb some of the bright sunlight from the skylights on the top floor of the building. His laboratory incorporated 'witches' cauldrons' of steam (the steam was designed to emulate the warm, moist internal environment of warm-blooded bodies, as bioreactors do today).[19] His later work in support of theories of eugenics also speaks to an aesthetic sensibility; eugenics operates within an aesthetic framework above all else, producing a uniform society free of anything unsightly. Ionat Zurr locates Carrel's aesthetic in his process: 'the ontological questions thrown up by Carrel's scientific experiments ironically resulted in his mystic and eugenic tendencies'.[20] Zurr links the dark steamy depths of his laboratory with his later work in eugenics, and speaks to how tissue culture practices shift understandings of life itself.

Carrel's 'elaborate rituals' – steaming the rooms warm and spraying the laboratory walls and floor with water, dressing surgeons in black gowns and flaming the narrow mouths of specially designed flasks – have been replaced by bioreactors and laminar flow hoods which direct filtered air over the tissue culture workspace. Spraying the room certainly would have reduced the number of bacteria floating in the air, but the hood is a huge improvement: airflow minimises contamination from the outside environment. Laminar flow hoods are typically stainless steel and brightly lit, with a set of UV light bulbs available to sterilise the entire workspace. Carrel's laboratory inverts this reflective brightness, offering itself as both a physical model of the warm-blooded bodily interior and a backdrop for breaking open and illuminating the same. Landecker writes that:

> The individual body of the experimental animal – one with an interior and an exterior, and opaque solid that was dark on the inside – was with the advent of tissue cultures supplemented in the laboratory by an interior set of life processes that had been extracted, distributed, and persuaded to live outside the body, glass-enclosed but always in full view.[21]

By creating a dark, steamy, interior on a large scale, Carrel's laboratory acts as an intermediate step between the body and the world – a dark, safe enclosure where tissues consider the possibility of emerging into the harsher light beyond.

A 2007 book about Carrel presents a parallel exploration of his work with eugenics and his 'secret' work with Charles Lindbergh pursuing research that might lead to human immortality. Provocatively titled *The Immortalists: Charles Lindbergh, Dr Alexis Carrel, and Their Daring Quest to Live Forever*, David Friedman has both written a biography of Lindbergh and Carrel and reread Carrel's work to emphasise a controversial 'immortality' interpretation. Friedman proposes that Carrel and Lindbergh's perfusion experiments had a 'secret agenda': the mechanisation of human bodies, producing a system where the body, like an airplane engine, could be augmented with new parts, thus prolonging life, perhaps indefinitely. While Malinin and Landecker hold to the theory that perfusion experiments were pursued in order to study organs in their living state, Friedman suggests that the 'glass heart' invented by Lindbergh and used by Carrel in his experiments was part of a larger plan to

> create a glass chamber of life – a place where whole organs could live, fully functioning, outside the body in which they'd been formed. These organs could be kept alive indefinitely, and, if Carrel was right, could be used as replacement parts to keep a human being alive. Maybe even forever.[22]

Friedman's book illustrates how the language of 'immortality', whether applied to cell lines or certain privileged human beings, was bound to 'the tissue culture point of view' in the United States in the 1930s and 1940s.

Humans were not the only animals to be immortalised through tissue culture technologies. An extensive literature of speculation about animal-based artificial meat products, their perpetuity and their possible manufacture dates back to speculative writing of the 1920s and 1930s, indicating that tissue culture has a history of inspiring creative work. Julian Huxley's short story 'The tissue culture king' (1927) combines ancestor worship with domestic tissue culture projects, creating a world populated by a tribe of bio-engineered human freaks living in households that are kept busy maintaining the cells of their ancestors and chief *in vitro*. Winston Churchill's essay about the future published in *Popular Mechanics*, 'Fifty years hence' (1932), predicted engineered chicken meat. Today, artificial meat survives in science fiction and research. Margaret Atwood's *Oryx and Crake* (2003) includes a lab stocked with the chicken-breast producing quasi-organisms imagined by Churchill, as well as organ-rich pigs ('pigoons') engineered and bred for the express purpose of organ transplant. These three narratives all provoke new ways of physically interfacing with meat. All three imagine futures where we eat differently: the bioreactor becomes a kitchen appliance.

'The tissue culture king' first appeared in Hugo Gernsback's early science fiction magazine, *Amazing Stories – Scientifiction*. Huxley's short story embraces a discipline rarely addressed by science fiction of its time: biology. The story was most recently printed in James Gunn's series *The Road to Science Fiction*. In his introduction Gunn observes that biology was a popular theme for later writers, especially in the 1960s, but in the 1920s and 1930s biology was almost completely absent from the pages of science fiction magazines.[23] Huxley (Aldous Huxley's older brother) never published another story – instead, he focused on research and his work with UNESCO.

'The tissue culture king' follows an English explorer in Africa who encounters a tribe populated by 'a regular Barnum and Bailey show' of anatomical freaks.[24] After a close call with tribal authorities, we are introduced to Dr Hascombe, a white scientist captured by the tribe many years earlier. Hascombe relates his experiences with the tribe, describing how he proved himself useful to their systems of ancestor worship and reverence for their king when he demonstrated the visual structure of blood using a microscope. Hascombe convinces tribal elders to allow him to stay by delighting them with the visual differences between human and bird blood. He secures his place in their society by creating giants, dwarves and obese virgins after tampering with their glandular systems. His tissue culture projects, housed in the 'Institute of Religious Tissue Culture', reproduce the cells of the king and the ancestors of the king's subjects, creating semi-living tissue samples that are archived in the institute and kept privately in the homes of the king's subjects.[25] Maintaining these cultures supplants the tribe's system of ancestor worship, enabling older members of the household to lead freer lives, as their presence is no longer required during rituals. Their extended bodies take the place of their living bodies.

Huxley introduces the term 'subsidiary lives' to describe what the Tissue Culture and Art Project would have named the 'extended body' of the king.[26] 'Subsidiary lives' neatly combines the idea of the semi-living and the concept of the extended body, conveying the synecdochic quality of harvested tissue. Subsidiary lives require stewardship and care, and are a career boon for a tribal leader, whose interest in the king's subsidiary lives ensures him a prominent place in his government. Pieces of the king's extended body (and those of a family's ancestors) must be cared for, just as the semi-livings in the Tissue Culture and Art Project's installations must be fed and looked after. The worship of the living flesh of the king strengthens connection between religion and tissue culture, providing a rich backdrop for the Tissue Culture and Art Project's incorporation of ritual into feeding and killing cultured tissues.[27]

Huxley mentions Alexis Carrel in passing: the narrator relates a visit to Carrel's laboratory when touring the Institute of Religious Tissue Culture. His organised laboratory stands in for scientific power and influence. The story's concern with biology in general and Carrel more specifically brackets a larger question Huxley asks repeatedly: what responsibilities do scientists have and how should they use their knowledge to contribute to, or even engineer, society? Having escaped from Africa and returned to England through a series of incredible events, the narrator relates his experiences in Africa as he approaches the end of his life. Reflecting on Dr Hascombe's work with the tribe, he asks 'what end did all this power serve?', proposing two paths for scientists: 'those who labour because they like power, or because they want to find the truth about how things work'.[28] The narrator puts Hascombe in the former group, someone who could not resist remaining at the head of the society he created, implying that the narrator himself labours because he wants to 'find the truth about how things work'; after all, he did return to England and resume his career rather than remain a social architect of an isolated, 'primitive' tribe.

Hascombe's compulsion to remain powerful and indispensable in Africa rather than return to England and give up his colonial dominance speaks volumes about how race determines the setting and subject of the story. In the world of Huxley's story, would an industrial Western society be caught in the throes of ancestor worship, with the desire for obese virgins used for ritual purposes? This possibility seems most unlikely, given the colonial attitudes of Dr Hascombe and the clear separation between civilised Europe and primitive Africa. 'The tissue culture king' was not the first and would not be the last tissue culture narrative to be complicated by race and racism. Racism *in vitro* reached a discursive climax 30 years after this story was published with the decidedly nonfictional HeLa controversy. The HeLa cell line originates from the cervical cancer of Henrietta Lacks, and was harvested without her knowledge or permission in 1951. HeLa was the first 'immortal' cell line: HeLa cells survive multiple divisions without dying. HeLa cells went on to have a distinguished career in medicine. Among a long list of research advances facilitated by Lacks's cells, highlights include contributions to the polio vaccine and being the first human cells to be cloned. Despite these huge breakthroughs, this particular cancer was found to be extremely aggressive, capable of contaminating other cell lines. Of all the possible reasons for this contaminating behaviour, it was Lacks's race that ended up at the centre of the debate; her blackness was what was really contaminating.[29]

'The tissue culture king' does not explicitly suggest that blackness is contaminating or dangerous, but the story's basic premise anticipates the HeLa controversies. Although the king does consent to his cell biopsy, the story describes the unwillingness of elder ancestors to submit to harvesting, even referring to a law that gave children aged 25 and older 'complete control over them'.[30] Similarly, the genetically manipulated children had no way of consenting to become bearded ladies, giants, dwarves or obese virgins bound to religious vocations. HeLa cells were propagated without Lacks's consent in much the same way that ancestor cells were collected and archived at the institute. The tribal people of 'The tissue culture king' demonstrate infectious, contagious behaviours that enable them to be, for example, hypnotised en masse. This group hypnotisation sets the stage for the story's final scene, where Hascombe, under the influence of his own hypnosis technology, turns back from an escape with the narrator to continue his work in Africa. The ability of these two white men to control the actions of an entire society (protected from their own technology by a twin set of tin foil hats) turns black bodies into raw biomass, subject to the '*zoe*-philic' desires of white men, equivalent to the cell tissues Hascombe has been manipulating throughout the story.[31] This easy transfer of behaviour from one person to another recalls HeLa's rapid contamination of other cell lines and, again, the racial rhetoric behind these behaviours echoes the 'one drop' theory of racial contamination. In the world of 'The tissue culture king', tissue culture and genetic manipulation signal inevitable, irreversible technological domination: the only way out is running away.

Margaret Atwood's novel *Oryx and Crake* (2003) introduces a post-apocalyptic landscape inhabited by feral laboratory animals; a human society made up of the Children of Crake, human beings created and engineered by the scientist Crake (formerly Glenn); and Snowman (formerly Jimmy), Crake's childhood friend. As with the African tribe in 'The tissue culture king', the Children of Crake are primitive, suggestible and easily coerced into activities that closely resemble the mass hypnosis Hascombe experimented with in Africa. For example, they believe that Snowman has a direct line of communication with Crake, and obey whatever communications Snowman receives. Although they had been designed to forgo religious inclinations, the women worship the deceased Oryx, Crake's lover and caretaker of the children. Children of Crake perform various tasks for Snowman, including preparing his fish dinner (they are vegetarian, with ruminant digestive systems based on those of rabbits).

Through a series of 'replays', we learn how Snowman arrived at his current situation. Snowman first recalls his life as Jimmy, a 'good boy', whose mother abandons him at around the time Crake arrives. His father works for biotechnology companies tasked with developing animals like pigoons, which serve changing purposes. Initially designed as living organ farms, pigoons are repurposed to host organs for transplant into humans. The second company Jimmy's father works for, NooSkins, uses pigoons to develop transplantable skin grafts and as hosts for human neocortex tissue. As a side hobby, the researchers create hybrid animals, like Jimmy's pet Killer, a rakunk (raccoon and skunk). Mild mannered and without the smell of a skunk, Killer departs with Jimmy's mother, who 'liberates' his pet before smashing her own and her husband's computers. Company security suspects her of being part of an ecological guerrilla organisation working to undermine the research Jimmy's father carries out at NooSkins.

Food has also been affected by these future biotechnologies. While visiting Crake at his exclusive job with Watson-Crick laboratories, Jimmy responds with visceral disgust when introduced to the 'animal-protein tuber' that would be named the ChickieNob.[32] The new incarnation of chicken 'was a large bulblike object that seemed to be covered with stippled whitish-yellow skin. Out of it came twenty thick fleshy tubes, and at the end of each tube another bulb was growing.'[33] Although Jimmy isn't thrilled by the idea of eating a ChickieNob product, later in the book he eats ChickieNob 'Nubbins' without thinking twice.

Winston Churchill predicted Atwood's ChickieNob with his 1932 essay about the future, 'Fifty years hence'. Published in the United States in *Popular Mechanics*, Churchill writes:

> We shall escape the absurdity of growing a whole chicken in order to eat the breast or wing, by growing these parts separately under a synthetic medium. Synthetic food will, of course, also be used in the future. Nor need the pleasures of the table be banished. That gloomy Utopia of tabloid meals need never be invaded. The new foods will be practically indistinguishable from the natural products from the outset, and any changes will be so gradual as to escape observation.[34]

Animal rights activists and proponents of artificial meat have taken up this passage of 'Fifty years hence'. Atwood conveys the tension between creating a monstrous living thing like the ChickieNob and advocating for its rights. For example, when God's Gardeners, an eco-guerrilla organisation, 'liberate' the immobile fleshy blobs during the end days of Crake's pharmaceutical extinction, this activity questions the meaning of 'liberation' and the boundaries of autonomous organisms. Can ChickieNobs be free when their bodies are closer to vegetable than animal? This tension resonates with the Tissue Culture and Art Project's concern that semi-living entities could become a new class subject to exploitation. The exploitation or liberation of a ChickieNob is a murky ethical area that Atwood maps by describing media responses to ecological guerrillas setting legless, nearly inert piles of food-production flesh 'free'. What might it mean to 'liberate' Carrel's immortal chicken heart, or even a contemporary broiler chicken, bred to be almost as immobile as a ChickieNob?

Atwood examines the subtle, even sly transition Churchill speaks of by tracking Jimmy's changing appetites. As ChickieNobs penetrate the fast food market more deeply, Jimmy buys Buckets O'Nubbins (the popular product made from ChickieNobs) with regularity.[35] The process of transitioning from chicken to ChickieNobs is a process of forgetting: forgetting the origins of the food, and forgetting the changes that have precipitated the engineered food's existence. In Atwood's words, 'forget everything you know about the provenance'.[36] ChickieNobs interface with tissue culture meat both in terms of cognition (these beakless-brainless-legless creatures must be some form of semi-living) and in terms of wholeness. While the extended body of a tissue sample semi-lives in growth medium, its mass increasing irrespective of its original size, the ChickieNob is an animal in itself, without the synechdochic qualities of tissue culture experiments. Yet there is something partial and unformed about the ChickieNob. Buckets O'Nubbins are not individual parts or even series of parts extracted from a whole. The ChickieNob was never whole; this semi-living organism was born incomplete. Reduced to its useful parts, as Churchill describes, the ChickieNob escapes one absurdity (that of a real chicken) for another (that of a collection of semi-living parts that do not add up to a whole, instead existing somewhere between raw biomass and organism).

ChickieNobs are not the only hybrid species to be released from captivity. When observing 'wolvogs', dogs engineered to attack while appearing benign, Jimmy asks Crake whether they might escape and 'spiral out of control – like those big green rabbits'.[37] Eventually wolvogs do spiral out of control, becoming a major threat to Snowman in the forest where he lives with the Children after the world is depopulated in a pharmaceutical extinction engineered by Crake. During this devastating plague, the ecological guerrilla group God's Gardeners liberates wolvogs and pigoons along with the hapless ChickieNobs. Pigoons can be as aggressive as wolvogs, as Snowman finds out later in the forest. By first suggesting that the relatively benign glowing rabbit might escape and interbreed with the general rabbit population, Atwood's later liberation of speculative species like wolvogs and pigoons capitalises on our objections to creatures like GFP rabbits being 'free' to leave their laboratory homes. Playing on the collapse of laboratory and domestic space explored by artist Eduardo Kac and his 'Free Alba' campaign, Atwood offers the harmless escape of rakunk Killer from Jimmy's company-owned home alongside the violent liberation of more grotesque species not designed for domestic interaction. Glowing rabbits, descendants of Alba the *GFP Bunny*, straddle domesticity and wildness: lab animals have created a new wilderness.

Although Crake wants to remove the 'God gene' from his Children, they nevertheless develop a cosmology that embraces voices from the sky, spirits and the compulsion to perform elaborate rituals (Snowman's fish, for example). Some of their actions are circumscribed by their genetic programming (such as marking territory with urine) but others, like the fish procession, develop independently of Crake's intentions when building them. Many of the book's insights involve food and new pleasures of the table. Crake gave his Children digestive systems that resemble those of rabbits, requiring a diet of leaves and fresh greens, which are then recycled. Jimmy and the Children have an awkward moment when they present him with some of their best caecotrophs, 'semi-digested herbage, discharged through the anus and re-swallowed two or three times a week'.[38] Caecetrophs compel the Crakers to sustain their food in their own bodies. Their digestive systems open up food spaces that stretch even Jimmy's imagination.

Snowman spends a good deal of time considering food, both while living comfortably as Jimmy and when trying to simply survive, post-apocalypse, as Snowman. Food reflects the massive changes happening in his environment and his methods of adapting to those changes. When living among the Children of Crake, their fish procession has the most ritualised valence of all the food moments in the book, suggesting that even for people who consume caecetrophs, 'eating is never a "purely biological" activity'.[39] Although he never tastes the Children's caecetroph gifts, Jimmy's gradual acclimatisation to ChickieNobs and the constant, casual introduction of unusual food products help make taste the most vivid sense in the novel. Jimmy and Snowman taste technology. While the Tissue Culture and Art Project's installation *Disembodied Cuisine* attempts one jarring, ambitious leap into cultured meat, *Oryx and Crake* describes the seamless, subtle process Churchill anticipates in 'Fifty years hence'. Jimmy can still appreciate 'realness', and immediately recognises the superior table laid out at Watson-Crick when visiting Crake, but he has also adapted to Nubbins, soy replacements for meat and other engineered food. Although *Oryx and Crake* begins after the transition from traditional to *in vitro* meat has already begun, the novel is nonetheless a story of adaptation. As Snowman moves through changing food landscapes populated by semi-living tissues and accepts what he at first rejected, the transition to new pleasures of the table have begun.

But at what price? In her short essay 'Chicken', Donna Haraway reminds Atwood's readers that the speculative industrial poultry bodies of *Oryx and Crake* have no heads for a reason; they are 'tasty organs without organisms, especially without annoying heads that register pain and perhaps have ideas about what constitutes a proper domestic bird's life'.[40] The chicken's head allows people to relate to chickens as thinking, knowing entities. Chicken's supporters contrast her ability to recognise the facial features of other chickens and the pastoral activities of rural chickens (pecking, eating, preening) with her unfortunate life and death in factory farming (beak burned off, starving and thirsty, decapitated while still alive). Although beef and the industrial cattle complex are most often taken up as a context for artificial meat's desirability, chickens return again and again as living sites of genetic manipulation and environmental degradation. Churchill's predictions have come true; today's industrial chickens have been engineered for breast tissue, so much of it that broiler chickens often fall over or cannot walk in the first place.[41] Chickens develop extraordinarily rapidly: 'It's an impressive feat of agricultural engineering to produce a 4-pound bird on 8 pounds of feed in six weeks!'[42] Chickens and their eggs also enjoy a special place in genetically modified food and pharmaceutical research. For example, observe changes within the company Avigenics, where genetically modified poultry designed for consumption as food became a staging arena for pharmaceutical research. In 2000, their website outlined plans to 'change the poultry industry forever'. In 2008, their website emphasised 'bio-therapeutics for the treatment of cancer, infectious diseases, organ-dysfunction, genetic disorders and autoimmune diseases'.[43] Like Atwood's pigoons, Avigenics' birds are multipurpose animals.

These speculative futures point to an expanded understanding of Honor Fell's 'tissue culture point of view'. For Fell, the tissue culture point of view referred to several conceptual and practical aspects of tissue culture methodology, including changing meanings of death and life, the potential romance or mystique of this new technique, scaling effects that potentially link cells to entire animals and the time-based events of tissue culture that fascinated Carrel.[44] When tissue culture was a recent invention, the tissue culture point of view evoked a set of eyeglasses or lenses that allow for a changed way of seeing the world. As these biotechniques became more commonplace, the tissue culture point of view became more pervasive.

Tissue culture has thoroughly reordered our notions of living and dying: concepts like brain death and practices like organ donation are now fully integrated into the way we live and die. We do not need to adopt the tissue culture point of view; we are already part of it. Interspecies relationships, hybridity and scaling effects have also infiltrated our everyday interactions, with pharmaceuticals and genetically modified organisms demonstrating the capacity for organisms to take one another's place and even to merge across kingdoms. The speculative worlds of Atwood, Churchill and Huxley take the tissue culture point of view even further, proposing domestic spaces that have evolved to include extended bodies and new forms of livestock. Caring for and consuming these semi-living entities becomes a familiar part of daily life even as the lives of livestock and ancestors become subsidiary, the part standing in for the whole. Eating *in vitro* both compels diners to forget about provenance and also connects humans, machines and animals in patterns of recursive symmetry. The tissue culture point of view privileges the taste buds over vision. ChickieNobs become palatable, rabbit meat fluoresces and the bioreactor becomes a kitchen appliance.

3

Domestic Computing

Neiman Marcus Mandarin Orange Soufflé

Ingredients

1 ¼ c. orange juice (preferably from concentrate, thawed & diluted)
1 env. (1 tbsp.) unflavored gelatin
1 c. sugar
2 lg. egg yolks
1 ½ tbsp. fresh lemon juice
1 c. heavy cream, whipped to soft peaks
½ c. Mandarin orange sections (4-oz. can)

Instructions

1. To prepare the soufflés, pour a quarter-cup of orange juice into a small bowl, sprinkle the gelatin over, and stir to dissolve. Set aside to let the gelatin soften. Prepare an ice bath in a large bowl.
2. Pour the remaining orange juice into a small, heavy-bottomed saucepan, and stir in the sugar and egg yolks. Over medium heat, gradually bring the mixture to a simmer, stirring constantly until the mixture begins to steam and is slightly thickened. Do not allow mixture to boil.

3. Add the softened gelatin mixture (which will have a rubbery texture) and the lemon juice. Stir until incorporated, and transfer the 'custard' to a clean mixing bowl; sit the bowl in the ice bath to cool. While the custard is cooling, stir it occasionally.
4. Using a wire whisk or an electric whisk, whip the heavy cream until soft peaks form. With a spatula, gently fold some of the whipped cream into the cooled custard mixture to 'loosen' it, then add the rest of the cream mixture and fold in until fully incorporated.
5. Place three to four of the Mandarin orange sections in the bottom of six individual 5-oz. fluted plastic dessert moulds. Fill moulds with the orange soufflé mixture. Place the moulds on a cookie sheet and cover with plastic wrap. Transfer to the refrigerator and chill until firm, at least four hours or overnight.
6. Carefully unmold the soufflés, and place each one on a radicchio leaf. Place two endive leaves behind each soufflé and lean upwards. Arrange about half a cup of mixed fruit next to each soufflé, and place a large strawberry on top of the mixed fruit.[1]

~

In the 1950s and 1960s, a computer in the home became a realistic possibility. Machines were smaller, microprocessors were becoming more affordable and the popular imagination had been primed for domestic computers with machine-human hybrid characters like Rosie, the *Jetsons*' robot maid and the mechanical servants populating the comic book *Closer Than We Think*. Nearly all advances in domestic technology have started in the kitchen, with refrigerators, dishwashers and other appliances transforming housework. Computers would logically join these other appliances in the kitchen of the future. Today, many of the networked home functions that were first developed in the 1950s and 1960s are invisible, seamlessly integrated into 'smart home' systems and every day communications technologies. Domestic computing enlivens histories of food art and bioart by tracing a history of the kitchen as a laboratory and a technologised space. With domestic technology, our appliances become more than instruments of efficiency. They begin to acquire subjectivity, to keep us company, to join us at the table as companions.

Outside of the home, women performed the clerical task of programming. Women were believed to have the dexterity and sensitivity necessary

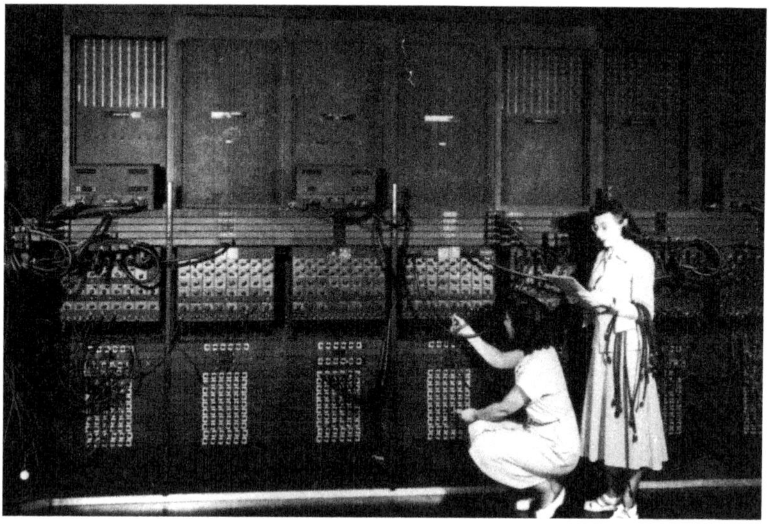

Figure 3.1 Marlyn Wescoff (standing) and Ruth Lichterman (crouching) wiring the right side of the ENIAC with a new program. U.S. Army Photo from the archives of the ARL Technical Library.

to manipulate delicate wires and mechanical parts.[2] Women programmers were themselves called 'computers'. In her study of the six women tasked with programming the ENIAC during World War II, Jennifer Light observes that while J. Presper Eckert and John W. Mauchly, ENIAC's developers, are now celebrated names in computer history, the six computers who programmed it, Kathleen McNulty, Frances Bilas, Betty Jean Jennings, Ruth Lichterman, Elizabeth Snyder and Marlyn Wescoff, were made invisible in their own lifetimes; they were even cropped out of press photos within days of the ENIAC's first public demonstration.[3] The joining of women and women's work with machinery and automation began with early computing projects like the ENIAC, entered the home with the Kitchen Computer and the ECHO IV, and continues today, with increasingly sophisticated and subtle interfaces. These new domestic technologies signal a transition in how we relate to machines. Joanna Zylinska articulates this transition as 'not so much a "human being" understood as a discrete and disembodied moral unity but rather a "human becoming": relational, co-emerging with technology, materially implicated in sociocultural networks, and kin to other life forms', including mechanical life forms.[4]

The first domestic computer prototypes both emulated and excluded women's labour – sometimes serving as prostheses for homemakers, at other times collapsing feminine and masculine spheres in absurd ways. Early efforts to bring computers into the home revealed the paradoxical ways in which women's work and women's bodies were conflated with machinery, as with the ENIAC 'computers', and the ways in which advertising and popular imaginings of domestic technologies further develop 'a growing awareness among women of how the home shapes female identity'.[5] Home Economics transformed the recipe with standard measurements and tutorials on technique. This transformation allowed recipes to be perceived as rational objects of study that could be manipulated by computer scientists and programmers (we continue to describe computer language manuals as 'cookbooks'). For early home computer developers, recipes were problems to be solved, data points to be sorted and manipulated.

The first computers designed for use in the home tackle the 'problem' of women's labour by taking recipe storage and meal planning as proxies for a range of housework tasks. Alongside more recent consumer products, the Sutherland family's Electronic Computer for Home Operation (ECHO) IV (1966) and Honeywell's Kitchen Computer (1969) write a history of how domestic chores, recipe storage and meal planning became automated, and how automation became networked and responsive. Theoretically, the large boxes housing early domestic computers mark the kitchen as a space of technology and experimentation. In practice, these two machines functioned in dramatically different ways. The Kitchen Computer was more or less useless to the average homemaker – the machine addressed conceptual problems like meal planning, ignoring the more demanding household tasks. Removing the conceptual work from a homemaker's plate positions the housewife as an ambivalent manager. The conceptual divestment that accompanied machines like the Kitchen Computer contributed to fears that domestic machines would eventually usurp the homemaker's role – a fear that is quickly assuaged when women contribute to domestic computing, as with the ECHO IV.

More than a kitchen appliance, the ECHO IV taps into the regulatory systems of the entire home. Extending from the basement into every room of the house, the computer adjusted the television antenna to optimise reception, tracked expenses, regulated the temperature and kept the household to a standard time, with binary coded decimal clocks in every room. Unlike the theoretical users of the Kitchen Computer, who learned to communicate with the machine in order to access predefined programs, Ruth Sutherland had an active role in programming the ECHO IV.

The ECHO IV and the Kitchen Computer addressed many of the domestic tasks Home Economics had been working to streamline and quantify. Both computers were designed to assist homemakers with recipe storage, grocery shopping and meal planning. At the time of their development in the 1960s, the concepts behind these machines seemed futuristic, the stuff of science fiction, but the activities they were designed to perform are integrated into twenty-first-century homes, as 'smart homes' and internet recipe databases become increasingly prevalent. At their abstract, conceptual best, these projects could be considered technological fulfilments of Home Economics principles, giving women more efficient ways to organise their homes from the clean, distant perspective of the programmer. The Sutherlands presented the ECHO IV to Home Economics classrooms and organisations, indicating that home economists welcomed (or were at least curious about) technologies like the ECHO IV.[6] Yet, as a field, Home Economics did not expand into Computer Science. At the corporate level, men continued to create and market domestic machines with little regard for the real constraints faced by women working in the home. As Judy Wajcman observes in her book *Feminism Confronts Technology*, 'women have not been the prime beneficiaries of domestic technology', which is controlled and 'designed by men in their capacity as scientists and engineers, people remote from the domestic tasks involved'.[7] This distance from domesticity marks the Kitchen Computer and the ECHO IV as contrasting prototypes not only of future smart homes but also of gender relations and women's labour in domestic spaces.

Created by Westinghouse Electric engineer Jim Sutherland, the ECHO IV began life as leftover parts from surplus Westinghouse controller hardware. Writing in *IEEE Annals of the History of Computing*, James Tomayko observes that 'newspapers and magazines mostly tried to measure the impact of the machine on the family, rather than make any technical evaluations'.[8] A lack of interest in technical evaluation should not be too surprising, given that the ECHO IV was the first of its kind and could not be evaluated against any other domestic computers. Concern with 'impact

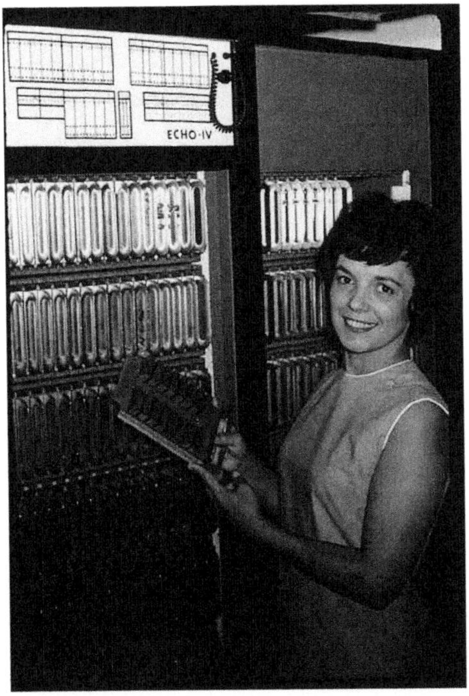

Figure 3.2 Ruth Sutherland at home with the ECHO IV. Photo: Jim Sutherland.

on the family' confirms that even the very earliest home computers necessarily engage the family dynamic and contribute to the household. The ECHO IV was featured in the April 1968 issue of *Popular Mechanics* with the incredulous title 'A computer in the *basement?*'[9] Currently residing in the Computer History Museum's off-site storage in the NASA/Ames Research Facility in Mountain View, California, the ECHO IV gathers dust alongside a collection of other old computers, its components still housed in their original wooden cabinetry. In addition to its two large input and processing units, the ECHO IV incorporated a series of panels throughout the home, which communicate with the device. The ECHO IV lived in the basement, where it took up a substantial amount of space (it is about the size of two large bookshelves) and produced a significant amount of heat. The *Popular Mechanics* article focuses on Ruth Sutherland's active role in designing the computer, introducing her with the question, 'Will it replace *me*?'[10] The spectre of a mechanical bride signals the collapse of household machines with housewives, their bodies and their labour.[11] This simultaneous fantasy and anxiety about technology replacing women's work and bodies has been seen before in rhetoric surrounding the washing machine, the dishwasher and other household appliances.[12] These devices do household chores with a side helping of sexual and libidinal satisfaction, standing in as objects of love for the men who design and then purchase them. For the Sutherlands, the mechanical bride need not be feared: Ruth Sutherland 'isn't worried about [being replaced] now, since she has learned that *home* computer programs must first be flow-charted by someone who knows *homemaking*'.[13] The Sutherlands model a method of engagement that will return later in this volume with bioart practitioners who call themselves 'public amateurs'. By learning in public and valuing do-it-yourself approaches, the Sutherlands show how some of the problems that plagued machines like the Kitchen Computer might be avoided.

While the ECHO IV was a private experiment in the Sutherlands' suburban home, the Kitchen Computer was theoretically available for purchase (though no sales records exist). Now part of the permanent collection of the Computer History Museum in Mountain View, California, the Kitchen Computer was first shown in the 1969 Neiman Marcus holiday catalogue. A Honeywell minicomputer has been repackaged in a sleek shell that incorporates a counter surface and a chair. The catalogue illustration shows a woman wearing a stylish apron; she is ready to cook, with fresh fruit and vegetables surrounding her computer. She adopts a graceful spokes model pose as she basks in her high-tech kitchen. The Kitchen Computer retailed for $10,600, which included an apron as well as a two-week programming course. The text accompanying Neiman Marcus' advertisement reads:

If she can only cook as well as Honeywell can compute.

> Her soufflés are supreme, her meal planning a challenge? She's what the Honeywell people had in mind when they devised our Kitchen Computer. She'll learn to program it with a cross-reference to her favorite recipes by N-M's own Helen Corbitt. Then by simply pushing a few buttons obtain a complete menu organized around the entrée. And if she pales at reckoning her lunch tab, she can program it to balance the family checkbook.[14]

Addressed to techno-forward husbands, the advertising copy suggests contradictory readings. The woman is both aided and undermined by this computer. She is the pinnacle of Taylorist Home Economics principles, maximally efficient thanks to new technologies and therefore powerful, but she is at the same time somehow incompetent at meal planning and financial management, implying a lack of conceptual ability. She is not the computer's equal ('If only [...]'), but she has mastered soufflé (the recipe that opens this chapter might have been programmed into the Kitchen Computer's database). Unlike Ruth Sutherland, this spokesmodel has not contributed to the computer's design.

The Computer History Museum catalogue describes the Kitchen Computer as a playful mistake, highlighting the improbability of a housewife operating this machine ('Does 0011101000111001 mean broccoli? Or carrots?').[15] Unlike today's intuitive graphical user interfaces, the machine used labour-intensive binary code (a now defunct programming language called BACK) and without an accompanying teletype machine, its user interface consisted only of lights and switches.[16] The effort required to program and communicate with this machine makes using it for everyday tasks impractical. The possibility that a housewife would take this programming course collapses masculine and feminine spheres, trusting the housewife with esoteric knowledge that would challenge the simple expectations the advertising copy proposes for women in the home. Cowan traces 'the doctrine of separate spheres' to pre-Industrial Revolution changes in the conceptualisation of 'home'.[17] Until the 1960s and the feminist movement, the separation of spheres was a normal state of affairs to be capitalised on by household technology industries. The idea that a woman would dwell in the masculine sphere of high technology in order to better manage her kitchen inverts the entire system, carving out space for a utopian relationship between women, technology and homemaking. The Kitchen Computer reaches for parity even as it implies that collapsing these two realms would be more absurd than hopeful.

Valerie Aurora, LINUX developer and feminist, documented her visit to the Kitchen Computer with a photo essay. Her performance illustrates how ineffective household technologies can be when homemaker users are

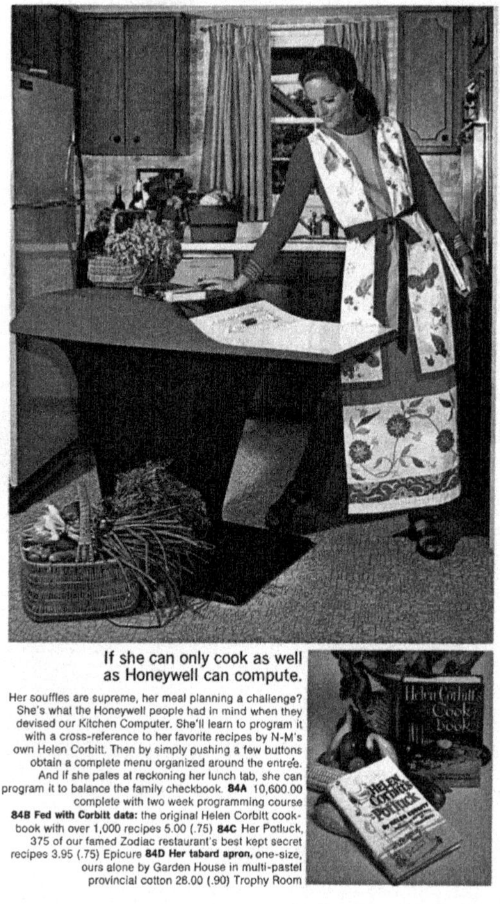

Figure 3.3 The Kitchen Computer, from the 1968 *Neiman Marcus Christmas Book*

left out of the design process. In 2008, when Aurora visited the machine, the Computer History Museum had not yet redesigned their exhibition space, and the Kitchen Computer was tucked into a purple cubicle, resting directly on the floor, level with visitors (its new display places it on a pedestal under dramatic lighting as part of an exhibition called 'Revolution: The First 2000 Years of Computing'). Using a series of photographs captioned with reflections from a 1960s homemaker character, Aurora's performance begins with excited curiosity about the machine, a gift from her husband. Soon she grows frustrated by the difficulties of using the machine. Eventually Aurora's housewife unleashes her feelings of rage, directed at the computer itself, her husband and computer engineering as a field. The last photograph depicts Aurora's performance character threatening the Kitchen Computer with a rolling pin: 'What a piece of junk! What idiot thought this was a good idea? I hate my Kitchen Computer! I hate my husband! I'm going to divorce him and go get a job designing computers that don't suck!'[18] By bringing the Kitchen Computer into dialogue with more contemporary attitudes about programming, usability and technology careers for women, Aurora enacts the real feelings that the imagined Kitchen Computer user might have felt, while simultaneously implying a second act, where things get really messy. The photographs evoke the opening scenes of Karen Finley's performances in the 1980s and 1990s, where Finley begins by performing normative feminine activities like rocking in a rocking chair or preparing food before erupting into uncontrollable rage and visceral food explosions. Aurora proposes a much needed intervention in domestic technology development: women should create and contribute to the machines designed to help them, killing the spectre of the mechanical bride and moving beyond the question 'Will the machine replace me?'

Neither the ECHO IV nor the Kitchen Computer could be considered explicitly feminist or fine art, but both enact early fantasies of melding domestic labour with technology, from two very different points of view. The ECHO IV is public amateurism at its best, fostering inclusivity, experimentation and participation from users like Ruth Sutherland. The Kitchen Computer, in contrast, does not support participation or experimentation from its homemaker users – preprogrammed corporate content overwrites whatever programming knowledge they might have gleaned from their two-week programming course. Despite the differences between participatory and corporate approaches, the machines have a lot in common. They both introduce unheard-of levels of automation and responsiveness to the kitchen. Food and eating are central. In the *Popular Mechanics* essay, the Sutherlands outline a plan to modify their kitchen cabinets so as to track the availability of food and allow the ECHO IV to assume responsibility for the shopping list. The Kitchen Computer theoretically allows its user to form meals based on available ingredients. Wajcman writes that 'gendered meanings are encoded in the design process' of domestic appliances, concluding that appliances continue to bind women to feminine spaces: 'domestic appliances "belong" in the kitchen, along with women'.[19] Ruth Sutherland's initial fear of being replaced by a computer reflects the fear of a sex/technology hybrid that could usurp both labour and pleasure.

Such fears arise from uneasy identification with machines, and the suspicion that machines perceive us as much as we perceive them. In her essay 'What do cyborgs eat?' Margaret Morse considers how 'smart' objects interface with people: 'the adjective *smart* attributes some degree of agency and, at times, of human subjectivity to the object world'.[20] Looking at objects and sensing or imagining that objects are looking back compels statements like Sutherland's: 'Will it replace *me*?' Features of today's 'smart home' appliances have been integrated into our perceptions of object-like behaviour to such a degree that most of the ECHO IV's automated tasks are seamlessly integrated and even invisible to twenty-first century consumers (timers on lights and thermostats, for example). A sinister side underscores these helpful appliances, summarised by Genevieve Bell, Mark Blythe, and Phoebe Sengers as the 'problematic design philosophy that devices, not users, should be in charge of users' activity'.[21] Smart appliances have both agency and control.

Recent smart appliances push the object into subjectivity more aggressively and visibly. In his short essay 'If you can't stand the coding, stay out of the kitchen', Dag Spicer compares the ECHO IV and the Kitchen Computer with a more recent smart appliance, the Electrolux Screenfridge (2000). The Screenfridge is essentially an internet-integrated refrigerator, including many of the features already incorporated into the ECHO IV and the Kitchen Computer (specialised menus based on available food and memos for family members, for example). The Screenfridge also offers barcode scanning, and places orders with your local grocer when items are running low. A more recent smart home innovation hacks a Nintendo Wii game console to control appliances, temperature, lighting and security, as well as audiovisual systems.[22] 3Com's 'Audrey' is a small computer shaped like a Tupperware box and designed to 'deliver an enjoyable Internet experience in the nerve center of the home. In most homes, that's the kitchen.'[23] Sutherland's ECHO IV, although not focused exclusively on kitchen tasks, kept a keyboard in the kitchen precisely because the kitchen is 'the nerve center of the home'.

These examples show some of the ways technologies have fulfilled the promises made by the Kitchen Computer and the ECHO IV. A computer running your home is no longer such a surprise – recall the astonished tone of *Popular Mechanics*' title for its coverage of the ECHO IV: 'A computer in the *basement*?' Continuing with the question 'who is in charge'?, smart appliances demonstrate surprising kinds of agency. Whirlpool advertises its remote monitoring systems as having a 'sixth sense'.[24] LG's 'HomeChat' lets you communicate with your appliances via text messages.[25] This communicative future has unintended consequences: for example, kitchen appliances like the Screenfridge have become outlets for spam malware. Domestic appliances join an expanding 'internet of things' made up of a growing network of largely unsecure devices and tracking systems.[26] For Anna Munster, smart devices bring with them a 'recursive creep [...] an everywhereness commonly called pervasive media'.[27] In the kitchen, pervasive media merge food, cooking, communicating and eating, creating a machine ecology where human–machine relationships' place becomes less and less clear, with humans less and less in charge. With domestic computers acting and networking on their own, our relationship to machines shifts from user to companion, enacting Zylinska's 'human becoming'. We share meals with our kitchen appliances every day, and our appliances may be more nourished by such encounters than we might imagine. Our constant companionship, our never eating alone, extends to include eating with lively machines.

4

Semiotics of the Kitchen: Feminist Food Art

Apron
Bowl
Chopper
Dish
Eggbeater
Fork
Grater
Hamburger press
Ice pick
Juicer
Knife
Ladle
Measuring implements
Nutcracker
Opener
Pan
Quart bottle
Rolling pin
Spoon
Tenderizer
U
V
W
X
Y
Z[1]

~

At the turn of the twenty-first century, the 100-year-long Home Economics project was reconfigured as critical art practice. Recipes become scores, orchestrating the performance of food. The recipe gathers a range of public and private discursive forms, melding the laboratory report with cooking practices handed down through generations. Marking a break from both domestic computing and Home Economics studies, second-wave feminist artists of the 1970s turned many assumptions from both of these fields upside down. Although the feminist movement shunned recipes and cookbooks, food and eating were central concerns. Feminist art responded to both Home Economics and domestic technology by developing tactics shared by food activists and ecofeminists. The kitchen became a political space capable of effecting change and mobilising resistance. To trace the kitchen's transformation, I focus on Martha Rosler's *Semiotics of the Kitchen* and related kitchen performance videos, Judy Chicago's *The Dinner Party* and complementary works by Suzanne Lacy, the collectively produced *Womanhouse,* and Linda Montano's 'Food & art' special feature in the journal *High Performance*.

Within a few years of the ECHO IV and the Kitchen Computer, feminist art intervened in the same domestic spaces these machines sought to optimise and control. Feminist artists were disinterested in or justifiably hostile to theories of organisation and Taylorist efficiency in the kitchen. Instead, their projects attend to the ways in which food and dining contribute to political mobilisation. In a special issue of *High Performance* (1981), performance artist Linda Montano gathers interviews and documentation from performance artists focused on food. She analyses the many ways that food appears in and activates art, finding that engagement with food manifests as 'political statement, as conceptual device, as life principle, as sculptural material, for nurturance and ritual, for props and irony, as a scare tactic, in autobiography, as feminist statement, in humor, for survival'.[2] All of the artists Montano interviews are feminist artists who produced work in the 1970s, and almost all of them continue to work today, sometimes extending and developing their performances over decades (for example, Alison Knowles regularly produces new performances of *Make a Salad*, first performed in 1962).

Second-wave feminist art insists on the public, political power of food and eating. In no other twentieth-century art movement were the radical ramifications of food, eating and the activity of preparing food so central to the reception and momentum of the work. When something as seemingly straightforward as a potluck presents itself as art, the ensuing emphasis on the critical capacities of activities like food preparation and presentation opens up space for thinking in multivalent, complicated ways about gender roles, industrial labour, transspecies encounter and agricultural policy.

Feminist art about food critiques Home Economics philosophies of domesticity and womanly virtue, while simultaneously (if unwittingly) sharing the progressive elements of Home Economics philosophy that promote new careers for women. Feminist artists of the 1970s politicise the personal and professional options open to women, often beginning with the artist's own choices. For example, when creating her monumental installation *The Dinner Party*, Judy Chicago articulates the play between art and political action by recounting her personal choices and the pressures she faced as a young artist. In a journal entry from 1977, Chicago writes: 'I am beginning to see that in taking on the Project I took on the real nexus of the problem that has prevented women from overcoming their oppression.'[3] She goes on to draw parallels between her studio, which was a workshop environment that reclaimed women's forms like china painting and embroidery, and the challenges women face in society as thinkers and artists.

In her book *But is it Art? The Spirit of Art as Activism*, Nina Felshin traces the roots of activist art practices to the new and challenging aesthetics of conceptual art that arose in the 1960s and 1970s. For Felshin, 'activist art represents a confluence of the aesthetic, socio-political, and technological impulses of the past twenty-five years or more that have attempted to challenge [...] the hierarchies traditionally defining the culture as represented by those in power'.[4] As we will see with contemporary recipe art, artists disrupt the unquestioned acceptance of industrial eating and genetic engineering, employing the same technologies used by these systems to subvert them. The impulses Felshin mentions animated the Home Economics arena as well: suggesting that a woman's private domestic processes might be scientific research implies that making such processes public disrupts norms and hierarchies. Domestic chores and tasks leave the home and are circulated in challenging new forms influenced by conceptual art and activist approaches to new political and technological discourses.

When I began my research, I assumed I would find many critical reinventions of the recipe among feminist artists. Shouldn't the recipe form lend itself easily to feminist projects? As traces of creativity and social exchange between women, recipes nurture knowledge systems that feminists would surely support. Furthermore, recipe development in Home Economics departments could be read as discursive community building. In the case of Cornell's department, historians of Home Economics

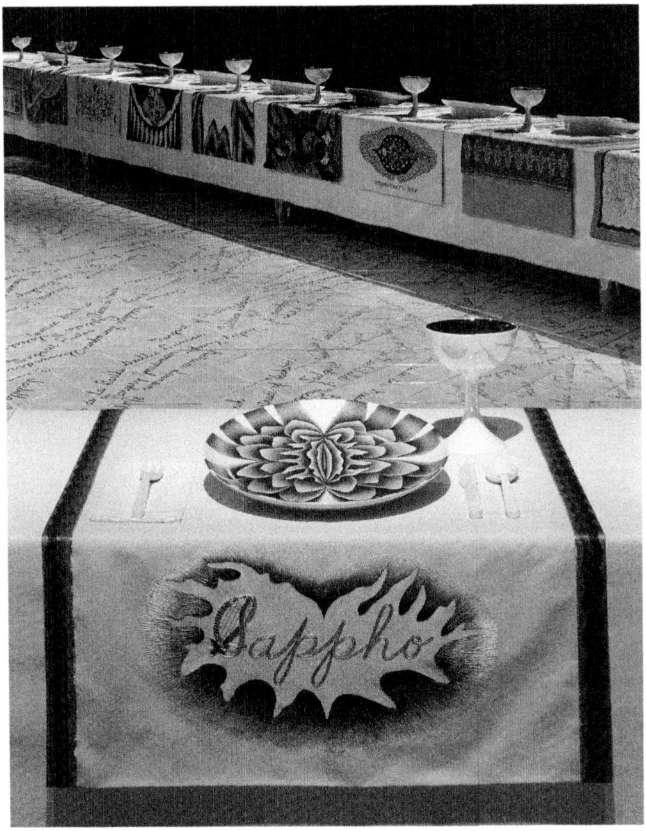

Figure 4.1 Judy Chicago, *The Dinner Party*, Sappho place setting (1979). Mixed media. Photo courtesy of Judy Chicago / Art Resource, NY.

credit the department with opening up feminist topics and dialogue within the university.[5] I was wrong; the recipe was not a common form among second-wave feminist artists. I underestimated the suspicion with which they regarded cookbooks, recipes and food preparation, all of which must have evoked the undesirable, even degrading, domestic worlds of many of their mothers and those of their peers who had not been liberated by the movement.

Given that I have learned a great deal about the political stakes of women's liberation while working in my mother's and grandmothers' kitchens, I was surprised to find that cookbooks and recipes were perceived as anti-feminist. Arlene Voski Avakian and Barbara Haber introduce their book *From Betty Crocker to Feminist Food Studies: Critical Perspectives on Women and Food* with this anecdote about how cookbooks and recipes were received in the 1970s:

> When Barbara Haber developed the large cookbook collection at the Schlesinger Library at Harvard University's Radcliffe Institute for Advanced Studies, many feminists and women's studies scholars were not supportive of her interest in these works. Cookbooks, they argued, were a mark of women's oppression and should not be collected in a major American library committed to the history of American women. The idea that cookbooks are documents of women's history, a perspective that seems so obvious now, was not generally accepted when Haber made this argument in the 1970s.[6]

This story about how Haber's research was received by feminist scholars (which goes on to describe an equally fraught response to her first book about women, food and cooking) suggests rhetorical and political reasons why feminist artists did not take up the cookbook and the recipe form in their work. Recipes and cookbooks symbolised the rigid social norms the feminist movement sought to overthrow. Feminist artists would cook and eat on their own terms.

Feminist art engaged food and cooking while also exploring instructions, measuring and scripted actions, but objections to archiving cookbooks betray their uncertain status within feminist circles. Perhaps cookbooks were seen as instruments of domestic control, or as participating in masculinist narratives that stressed culinary conformity and cooking as 'wifely duty'. My mother related to one of her first cookbooks, *Betty Crocker's Cooking for Two*, in precisely this way: she used the cookbook's menu planning suggestions every single night, composing meals that included complementary food groups, drinks and dessert, because this was what it meant to be a 'good wife'. At some point, cookbooks began to be valued as objects of cultural significance. My mother looks back on her *Cooking for Two* days with the same critical eye that we bring to recipes and cookbooks today, acknowledging that cookbooks produce behaviours and roles like 'wife' and 'mother', even prescribing what wives and mothers should do in the kitchen and how they should relate to their families. What brought about this shift from passive obedience to critical self-awareness? Many feminist art projects appropriate, reclaim and empower objects that had been perceived as oppressive and enslaving. Thanks to this systematic reclamation project, the feminist movement, even while shunning cookbooks, enabled critical readings and the eventual acceptance of these texts.

Take, for example, artist and *Womanhouse* collaborator Mira Schor's definition of 'recipe art' in her book *A Decade of Negative Thinking: Essays on Art, Politics, and Daily Life*. Appearing in a section called 'Trite tropes', Schor describes recipe art as the most formulaic of contemporary styles: 'the works that get the most attention, because their ingredients can be condensed into a provocative sound bite, are frequently the least interesting in person'.[7] Even as Schor equates recipes with the worst of the 'trite tropes', she offers a recipe of her own that supports the more generous, positive reading of recipe art that I develop in this book: 'Recipe: something from popular culture + something weird or expressive + useful promotional sound bite'.[8] Although the content of her recipe for recipe art has a negative connotation, using the recipe format demonstrates the playful and enduring qualities of the form. Echoing the Carl Pope poster that illustrates the chapter, the familiar recipe format conveys a conceptual idea.[9]

Although cookbooks continued to be perceived as markers of oppression, the recipe format flourished as a way of creating and documenting performances, and 'art recipes' circulated in FLUXUS and feminist circles.[10] Schor's recipe for recipe art inherits this mode. Bonnie Sherk's text works include recipes for *Conceptual Art* (1970) and for *Functional Art* (1980):

> RECIPE FOR CONCEPTUAL ART
> Just add water. Box of flour with light bulbs.
> RECIPE FOR FUNCTIONAL ART
> (⅓) Reason
> (⅓) Intuition
> (⅓) Passion
> Add two handfuls of humor. Blend well and mold. Set in motion and wait for synchronous tones.[11]

Merging ratios, sounds and abstract conceptual language, Sherk's recipes open up imaginary spaces that evoke our visualisations of finished food recipes. Feminist engagement with food was wide-ranging and focused on the politics of specific activities, like domestic labour, serving, dining and cooking. Sherk interrogates these roles with performances like *Public Lunch* (1971), 'in which [Sherk] ate a meal in the lion house at the San Francisco Zoo, adjacent to tigers and lions being fed raw meat', and *The Waitress* (1973–4), a performance character who served food in a range of contexts.[12] Collective meals, food and food service played a significant role in feminist art and political activity. Alongside Suzanne Lacy's consciousness-raising potlucks, artist-run restaurants such as Tina Girouard, Carol Gooden and Gordon Matta-Clark's FOOD and Allen Ruppersberg's Al's Café questioned capitalist assumptions within food service and food preparation and proposed radical alternatives that ranged from delicious to thoughtful but inedible.

Decades later, recipe artists continue to excavate the political content of feminist work, engage the FLUXUS idea of the recipe as script or score for performance and trouble traditional links between the recipe format and food preparation. This chapter's brief survey of feminist food art shows how food preparation and cooking became politicised within feminist art, establishing the ways in which food and food preparation could be politicised in other genres of practice. In addition to raising consciousness about women's roles in domestic spheres, this work stakes out space for resistance within previously innocent places, like the kitchen and the supermarket.

In Martha Rosler's video *Semiotics of the Kitchen* (1975), Rosler recites an alphabetised litany of kitchen tools while performing practical demonstrations that mimic and subvert the cooking demonstrations given in Home Economics classrooms. Of this performance, Rosler writes: 'A woman in a bare-bones kitchen demonstrating some hand tools and replacing their domesticated "meaning" with a lexicon of rage and frustration is an antipodean Julia Child.'[13] Although not a recipe in the traditional sense, *Semiotics of the Kitchen* performs the 'action' commands of recipes (chop, mix, blend, etc.). Rosler's performance is also a list, and a demonstration, adopting the look and feel of a distanciated, hostile and deadpan cooking show. Rosler disrupts expectations of any given item's appropriate use. The eggbeater clatters against an empty bowl, the knife cuts into thin air and the ice pick stabs aimlessly but violently at the butcher block counter. Rosler's 'lexicon of rage and frustration' lays bare the woman's

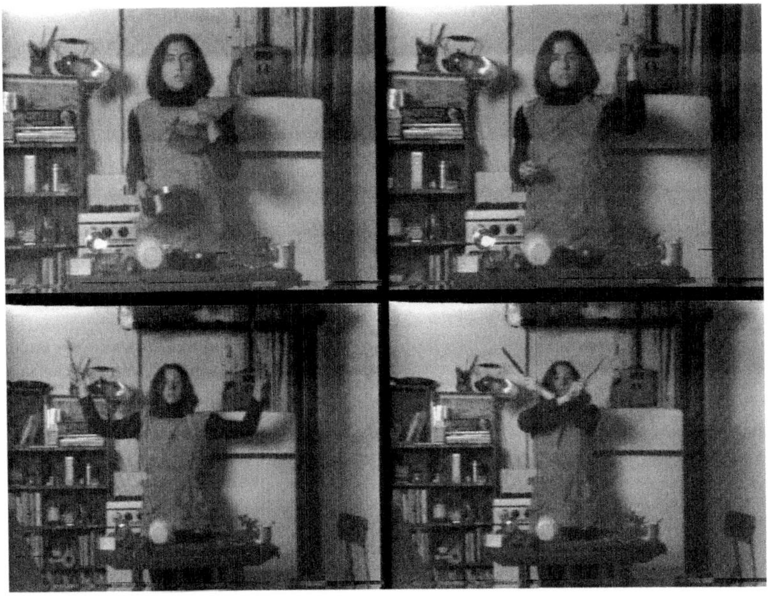

Figure 4.2 Martha Rosler, stills from *Semiotics of the Kitchen* (1975). Images copyright of the artist, courtesy of Video Data Bank, www.vdb.org.

confinement in the kitchen, offering limited, unsuccessful strategies for subterfuge and resistance.

We read the kitchen of the Kitchen Computer advertisement as a safe, unchallenging space, where a housewife joyously revels in her efficient, high-technology home of the future. Rosler's kitchen obliterates this complacency, replacing the spokesmodel's stylish home with a sense of almost post-apocalyptic doom. This kitchen has no food, boasts no shiny appliances and its unhappy occupant does not take pleasure in her work. Instead, each action channels her anger and illustrates her confinement, as if these tools construct not food but barriers and walls. *Semiotics of the Kitchen* makes the kitchen unsafe, opening it up as contested, political space. While present-day activist art challenges kitchen workers to resist consumer trends and GMOs, Rosler's empty kitchen focuses less on product and more on process, asking how the symbol systems and actions women are allowed to perform in the kitchen affect their political mobility.

Womanhouse, created in 1972 during Miriam Schapiro and Judy Chicago's time at CalArts, also critiques the kitchen. Located in suburban Los Angeles, the house was a performance and installation space constructed around collectively authored rooms. Victoria Hodgetts, Robin Weltsch, and Susan Frazier's *Nurturant Kitchen* features walls covered in three-dimensional round forms that vacillate between sunny side up eggs, breasts and a baroque wall covering. The dining room houses two- and three-dimensional representations of food but there are no chairs around the table, forcing visitors to experience the room as a wife or servant might, never able to sit down to eat. Unlike Rosler's kitchen, *Womanhouse* surrounds the visitor with food, though never real enough to taste. Instead, *Womanhouse* uses food symbolically, to explore relationships between women, food and food preparation, and the space of the house itself.

Margaret Morse writes of food as a 'symbol system': 'One could call both "eating"/"being eaten" and "enveloping"/"being enveloped" deep metaphors that pervade even the most advanced cultures and the highest art forms.'[14] For feminist artists engaged in the complex melding of cannibalism, narcissism and homo-eroticism, food operates symbolically and metaphorically, standing in for woman's desire go beyond feeding others and feed herself.[15] This symbolic transaction invokes incorporation more than ingestion or abstract identification. Morse distinguishes the two:

> Unlike identification, incorporation does not depend on likeness or similarity or mirrors in order to mistake the other as a the self; in an 'oral-sadistic' or 'cannibalistic' fantasy, the introjected object [...] is occluded and destroyed, only in order to be assimilated and to transform into its host.[16]

Womanhouse finds much of its power and controversy in the idea that symbolic food exchanges transform into the person incorporating the food, and that architecture might absorb both women and food. Returning to the *Nurturant Kitchen*, the house and the woman's body pull against and collapse into each other, evoking the hybridisation of woman and house seen in Louise Bourgeois's *Femme Maison* series. Of the drawing and sculpture series, which grafts women's torsos onto exteriors of free-standing houses, Bourgeois explains that the woman's assimilation into the house is unconscious: 'She does not know that she is half naked, and she does not know that she is hiding [...] she is totally self-defeating because she shows herself at the very moment she thinks she's hiding.'[17] Both *Femme Maison* and *Nurturant Kitchen* capture the tensions between introjected object and host Morse previously describes. The vulnerable body pulses within the walls of the family home, at once concealing and revealing a symbiotic relationship where woman and house are both compounded and antagonistic to one another.

Feminist artists symbolically nourish and eat themselves, incorporating food, its display and preparation, into acts of political resistance that transform visitors and artists alike. Feminist art eats itself; that is to say, the meal format ingests the connective tissue that binds femininity to food while simultaneously troubling the perceived 'natural' connection between women and food, producing a split consciousness of incorporation. In eating the self, one becomes both provider and consumer, simultaneously cannibalistic and narcissistic, both homo- and autoerotic. Food forms morph into body parts. Judy Chicago's *Dinner Party* invites diners to eat from plates that strongly resemble vaginal forms (although butterflies initially inspired the design concept).[18] *Nurturant Kitchen* promotes architectural identification, merging the body with the walls of the kitchen and transforming the bodies of visitors into fleshy extensions of those walls. With *Semiotics of the Kitchen* we eat from empty vessels, consuming our own rage and frustration. The question of what and whom we are feeding remains unresolved.

This symbolic nurturing and feeding does not make the rooms of *Womanhouse* or *Dinner Party* any safer than Rosler's kitchen. Although visitors are surrounded with objects to be eaten, the opposite effect takes hold; a kitchen that suffocates the visitor with food is almost as threatening as a barren kitchen bereft of anything edible. In the case of *Nurturant Kitchen*, food crawls up the walls and becomes embodied, giving the room breasts and rendering the walls fleshy. The room envelops the visitor even as she melds with the walls. Simultaneously enveloping and being enveloped, the eater folds into the eaten: 'the body of the other can be as large as an intrauterine-stomachic-intestinal interiority or virtual void within which one is "immersed"'.[19] Reading *Nurturant Kitchen* as such a bodily interiority (and as a 'virtual void', given that *Womanhouse* exists today only in memory) allows for an unsettling incorporation of the kitchen's excessive, smothering abundance. This unwilling engulfment can be extended to represent woman's position in the larger house, where she is similarly confined and swallowed whole. Being swallowed and engulfed evokes Judy Chicago's working title for what would become *The Dinner Party*: *Twenty-five Women Who Were Eaten Alive*.

The Dinner Party (1979) took five years to create, during which time Judy Chicago learned china-painting techniques and worked with hundreds of people. Taking the form of a monumental triangular dining table, each of its places sets the table for an important woman in history, with the names of more women written on the tiled 'Heritage Floor'. For nearly 30 years, *The Dinner Party* had no permanent home. In 2007, the installation moved to a permanent exhibition space at the Brooklyn Museum's Elizabeth A. Sackler Center for Feminist Art.[20] The early title, *Twenty-five Women Who Were Eaten Alive*, already suggests that these dinner plates serve up more than biographies or portraits of historical figures: the plates invite a cannibalistic act. We dine on the fleshy surfaces of history, relating to these plates as memorials, leftovers and monuments but also as bodies, erasures, digested matter: those who were eaten alive.[21]

Almost all of the plates on the table depict two- or three-dimensional forms morphing between butterflies and vaginal openings. While *Nurturant Kitchen* surrounds and envelopes visitors, *The Dinner Party* offers a memorial, an opportunity to witness the symbolic meeting of the women Chicago has invited to her table. The oversized cups and place settings combine with the size and monumental grandeur of the table to make visitors feel small. As with the dining room in *Womanhouse*, this is a metaphorical rather than visceral eating opportunity. Designed to be approached standing, *The Dinner Party* is not a sit-down meal, but its art exhibition context and monumental scale refuse to position visitors as wives or servants as with the dining room in *Womanhouse*. With a focus on the formal components and paraphernalia of dining, *The Dinner Party* takes dining as a metaphor, yet is not about food.

Suzanne Lacy hosted a parallel event when *The Dinner Party* was first exhibited called *The International Dinner Party*, where women all over the world dined together in honour of specific women of the past. Lacy comments on this project in an interview with Linda Montano:

> I've begun using food on a mass scale, partly as a result of Judy Chicago's *Dinner Party*. I have brought together women for massive pot lucks, probably because it's a metaphor for nurturing each other. When you look at what women do naturally, you

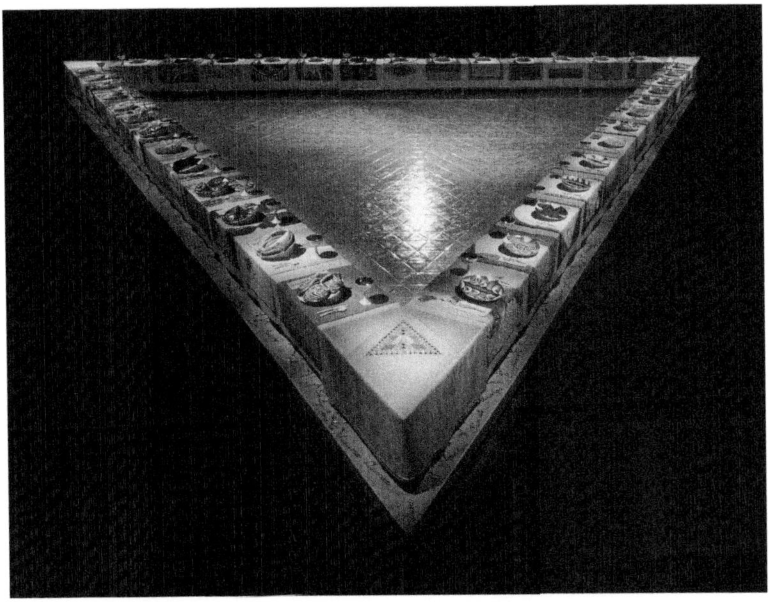

Figure 4.3 Judy Chicago, *The Dinner Party* (1979). Mixed media. 36 in. x 576 in. x 576 in. Photo courtesy of Judy Chicago / Art Resource, NY.

> find that they have made quilts or cooked or nurtured others. I've translated these ideas into art and have come up with large pot lucks [...] Both times I've done it, once with two hundred thirty women and another time with five hundred women, have been real powerful experiences. You just place all of this wonderful food on the table and watch the way the interaction happens. For the *International Dinner Party* I asked women to have simultaneous meals all over the world to celebrate Judy's show. Again it was sharing food to raise consciousness.[22]

Lacy describes a process of restaging and reinterpreting women's collective work (quilting, cooking, taking care). Participating in these activities as art, 'on a mass scale', and as political mobilisation interprets 'what women do naturally' without the prescriptive limitations of the Kitchen Computer or Taylorist models of housekeeping. Lacy brings food to Chicago's imposing table, focusing on conversation rather than consumption. Conversation ends up challenging potential essentialist outcomes of a phrase like 'what women do naturally'. In a consciousness-raising context, conversation marks food preparation and 'nurturing others' as critical, rather than innate or unconscious activities. The potluck network, too big to be contained in a private home, allows for self-reflexivity outside of the overdetermined context of the home and the standardising practices of Home Economics.

In Lacy's later work, food falls away but the dining table and conversation remain. Lacy produced *Whisper, the Waves, the Wind* in 1984. Women dressed in white occupy delicate white tables and chairs perched on the rocks and sand of two beaches in La Jolla, California. Their conversations about topics including beauty, sexuality, aging and death are broadcast to an audience gathered on the cliffs above. Judith Stein writes that Lacy 'adapt[s] the meal format'.[23] The 'meal format' connects this work, interested in the sensorial and conversational structures of the meal and the potluck, with art that takes the recipe as its format and space of adaptation. In Lacy's later performances, the dining table does not signify food and eating, but conversation and consciousness raising. Performances like the ambitious *Crystal Quilt* (1985–7) focus entirely on conversation, with tables and chairs citing the conventions of the dining room as a generative procedure rather than a foregone conclusion.[24]

All of these feminist projects about food and meals disrupt the assumptions that drive machines like the Kitchen Computer. Like Taylorist models of efficiency, machines such as the Kitchen Computer were produced in a cultural moment that viewed women as recipes view raw ingredients: chaotic systems in need of organisation. Evoking Lucile Ball's stint as a hyper-scheduled robotic housewife,[25] household computers presume an anatomical link between women and their homes, a very different sort of envelopment than in the living walls of *Nurturant Kitchen*, for example. Feminist art uncouples essential connections to remake women and food as voluntary interlocutors through performance. Moira Roth describes this process as a shift in context: 'Women artists have [...] addressed the topic of women and food, removing it from the realm of the merely personal in order to analyze the subject in societal and economic terms.'[26] Rosler shows that food preparation can be perceived as a set of neutral signs waiting to be inhabited by subversive figures like her 'antipodean' interpretation of iconic cook Julia Child, while *Womanhouse* and *The Dinner Party* reveal both subtle and obvious tensions in sites jointly occupied by food and domesticity.

All of these works remind us that kitchens and dining rooms are polluted, contested zones capable of complicating the meaning of domestic labour. Compared with the kitchens of home economists and the fantasy marketing materials advertising domestic computers, feminist kitchens are extravagant and sometimes even baroque. The kitchen emerges as an extension of the body, an architectural means of disgorging the private activities of women into public realms, and sites of political unrest, discord and empowerment. This group of performances and art projects concerned with the formal elements of dining and food preparation produces women who are not intrinsically bound to food and its preparation so much as they are historically and theatrically engaged with food and dining's semiotic potential.

5

DIY Coke

Coca-Cola

28 ml caffeine
28 ml vanilla extract
10 ml orange oil
10 ml lemon oil
10 ml nutmeg oil
10 ml cinnamon oil
10 ml koreander oil
10 ml neroli oil
224 ml alcohol
112 ml coca extract
84 ml lemon acid
224 ml lime juice
13.62 kg sugar
9.5 l water

Mix the caffeine and the lime juice in 224 ml boiling water. Add the vanilla extract, the orange oil, the lemon oil, the nutmeg oil, the cinnamon oil, the koreander oil, and the neroli oil when the mixture has cooled. Wait a couple of minutes and then add the rest of the ingredients and the water. Let the mixture rest for 24 hours.

Should make 10 liters of Coca-Cola.[1]

~

Consider how a box of cake mix maps fluid boundaries between the kitchen, the laboratory and the factory. Tutored by home economists, we forget the relationship between our individual cake's simple preparation and the industrial recipe created in the factory that packaged the mix. Factories, factory workers and industrial ingredients are often forgotten when consuming packaged foods. As feminist thinkers and makers begin to accept the recipe format, embracing cookbooks, kits and recipes as politically significant texts, industrial labour and manufacturing processes emerge as practices ripe for critical analysis. Understanding recipes and cookbooks as containers for critique rather than symbols of oppression has yielded what I call 'recipe art' – a mix of food art and feminist critique indebted to FLUXUS instruction works, scores and scripts. Recipe artists address the invisibility of industrial labour and corporate control of food supplies by bringing industrial recipes into the home to facilitate critical understandings of processed food. The kit format becomes an important vehicle for teaching and learning about both food and technology. In this chapter, I read Martha Rosler's *Romances of the Meal* (2000) alongside Maya Weinstein's *DIY High Fructose Corn Syrup* (2013). These projects peer into corporate research and development laboratories, bringing industrial recipes and laboratory protocols into our kitchens. What might we learn from preparing industrial food at home? How will reverse engineering top secret trademarked recipes deepen our understanding of culinary technoscience?

Martha Rosler's *Romances of the Meal* focuses on the architectures of industrial food production and the labour of women working in factory kitchens. Rosler recontextualises industrial ingredients and recipes, including reverse engineered recipes for the Coca-Cola formula (the actual recipe is a closely guarded industry secret).[2] In the *Laboratorium* catalogue, *Romances of the Meal* appears as pages of text with elaborate footnotes punctuated by history lessons and recipes. There are no pictures of the work or of the artist, or even of the food. Rosler focuses on chocolate and caffeine, including coffee and Coke, highlighting industrial baking recipes calling for hundreds of pounds of ingredients. Anyone who has eaten or worked to produce factory foods such as Twinkies and Coke is likely to be familiar with if not disgusted by these recipes. My mother once worked in a cookie factory; she still refuses to eat Keebler cookies. Rosler's *Romances of the Meal* evokes an analogous feeling of participatory revulsion. These recipes invite refusal. Their scale and technical complexity surpass the capacities of the domestic kitchen and the activities associated with their preparation are far removed from eating the finished product. Factory cooks never witness their food being eaten by people; instead, plastic and cardboard packaging materials devour the fruits of their labour.

Curators Hans Ulrich Obrist and Barbara Vanderlinden describe the *Romances of the Meal* archive and performance:

> Rosler investigates the industrial kitchen as a site of food production and as a metaphor for the channeling of creativity into things that are normalized, naturalized, and regularized, and whose creations are intended to be consumed rather than celebrated [...] The industrial kitchen [is] a place where there seems to be no accumulation and no apparent skill required, a place where women's creativity and inventions never develop.[3]

Like the kitchen of *Semiotics of the Kitchen*, the industrial kitchen leaves little space for comfort. The industrial kitchen is not a test kitchen; women working there do not conduct experiments or solve problems. Their work is predetermined, calibrated and routine. Their every step has already been choreographed and optimised by engineers: the industrial kitchen stages grotesque parodies of baking and cooking. Like Lucille Ball embodying the male engineer's fantasy of automated housework, the industrial baker's movements are scripted to be fast, efficient, repetitive. Obrist and Vanderlinden's observation that such work drains women's creativity implies that cooking should be different for women. Rosler has never shared this perspective; the 'antipodean Julia Child' of *Semiotics of the Kitchen* does not wait for creativity, invention or inspiration to strike.

Together, these recipes and archives critique how food, labour and industry intertwine. The written and performed components of *Romances of the Meal* trace an elaborate genealogy designed to unravel connections between factories, copyright holders, consumers and both industrial and domestic cooks. Insisting on textual encounters and bibliographic tracings, *Romances of the Meal* transforms the recipe into a vehicle for remembering and marking the provenance of ingredients. Here, foods like chocolate, coffee and Coke are given deep, thoughtful attention. Readers move from one to the next on a geographic and cultural tour, with 'romances' emerging in origin stories, secret recipes, travelogues and activist experiments designed to question assumptions underpinning existing food supply and distribution practices. The industrial kitchen exposes the standards of efficiency applied to domestic kitchens since Catherine Beecher and Christine Frederick began publishing efficiency studies in the mid-nineteenth and early twentieth centuries.[4] Home kitchens are more industrial than ever, with cooks using ingredients and processes borrowed from corporate research and development and techniques learned from test kitchens populated by corporate mascots.

After appearing in *Romances of the Meal*, Rosler's recipes were offered as 'artist instructions' on the e-flux website.[5] The result of collaboration between Christian Boltanski, Bertrand Lavier and Hans Ulrich Obrist, the 'DO-IT' project compiles artist instructions ranging from recipes contributed by Rosler and Rirkrit Tiravanija to a photo exchange from Christian Boltanski and storyboards from Matthew Barney. With each set of instructions, participants are invited to upload pictures and text documenting their 'realisation' of those instructions. Exhibitions have developed from the instructions, which function as scripts or scores, similar to those produced by FLUXUS artists.[6] I performed Rosler's recipe for homemade Coca-Cola at a 2007 event hosted by University of California Davis called Midnight University. Midnight University followed a daylong symposium devoted to the 'Lives of Documents'. In that context, the recipe as a travelling, lively document anchored the specific entanglements of the Coke recipe.

The Coca-Cola formula as a product of cooking, capitalism and consumerism invites countless rumours (for example, that two Coca-Cola company executives each know half the formula, so are consequently not allowed to travel on the same aeroplane). The secret formula has been discovered many times over in many contexts. The radio programme 'This American Life' devoted a show to reproducing and tasting one version of the recipe.[7] The Coca-Cola Museum in Atlanta, Georgia houses a theatrical secret vault where the formula could be viewed – if only visitors were

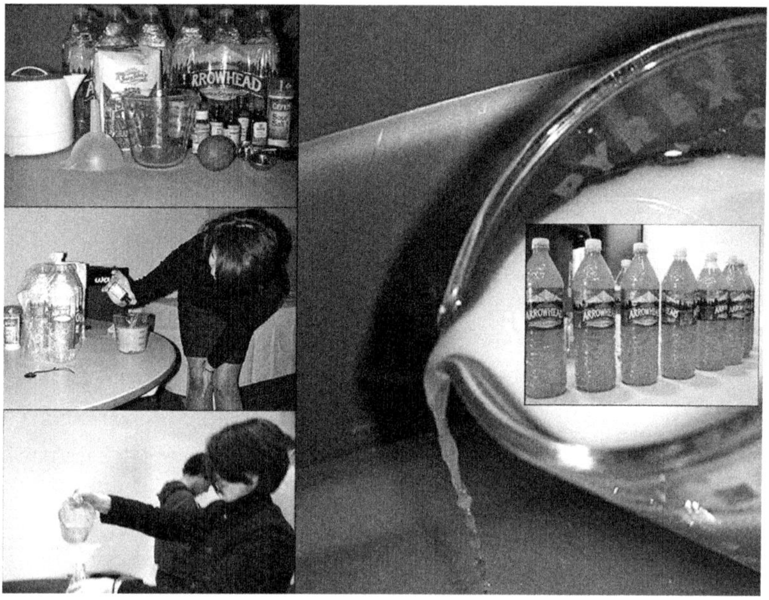

Figure 5.1 Lindsay Kelley, *DIY Coke* documentation (2007).

authorised to enter the stainless steel enclosure. Historian Mark Pedergrast published a version of the original formula in his book *For God, Country and Coca-Cola: The Definitive History of the Great American Soft Drink and the Company that Makes It.*[8] While Pedergrast's book was in press, Georgia antiques dealer Cliff Kluge offered a 1943 Coke formula for sale on eBay. In addition to recipes for Coca-Cola itself, countless recipes call for Coke as an ingredient. Rosler highlights Coca-Cola Roast, Chicken in Coca-Cola Sauce and Mexican Coca-Cola Muffins in *Romances of the Meal*. But reproducing the secret formula trumps these other engagements with Coke.

Over the course of several hours late in the evening during the Midnight University event, I bottled 10 litres of homemade Coca-Cola, using the recipe Rosler contributed to e-flux. Documentation from this event was then uploaded to e-flux as a realisation of Rosler's instructions. After completing this process, I reflected on my choice to use visual documentation. Uploading images seems antithetical to Rosler's text-only catalogue presentation of *Romances of the Meal*. Insisting on text promotes the idea that the artist produces more than aesthetic objects; she is also a historian, scholar and researcher.[9] Imagining text-based ways to document the Midnight University events led me to investigate other engagements with industrial kitchens and corporate recipes adapted for home preparation and eventually to write this chapter. *Romances of the Meal* brings industrial food ways into the home kitchen, critiquing the limits of factory and domestic food preparation. Performing and re-performing *Romances of the Meal* productively collapses industrial, domestic and institutional kitchens.

Turning from spaces of production to an individual ingredient, Maya Weinstein's *DIY High Fructose Corn Syrup* brings one of the most common industrial sweeteners into the home kitchen. Developed in the 1950s and 1960s as a cheap alternative to cane sugar, HFCS can be found in a wide range of packaged foods, from soft drinks to bread. HFCS begins with a high starch variety of corn, Yellow Dent. Yellow Dent corn must be processed before consumption as cornmeal, HFCS, taco shells and even biodegradable plastics. Like Yellow Dent corn, HFCS isn't available for purchase on its own – it exists only as an ingredient in other foods. Regular corn syrup may be found on supermarket shelves, but corn syrup and HFCS are different products entirely (although corn syrup may contain added HFCS). HFCS has converted a percentage of its glucose to fructose with enzymes developed especially for this process.

Homemade HFCS appeared in the 2007 film *King Corn*. Denied access to a HFCS production facility, Ian Cheney and Curt Ellis decide to make their own. After boiling corn for nearly two days and adding toxic ingredients like sulphuric acid, they taste their finished product, spit it out with a grimace and declare it 'sweet'.[10] Weinstein's kit makes Cheney and Ellis's experiment portable while also refining their recipe, sourcing difficult to find industrial ingredients and packaging everything in a wooden box complete with a USB-enabled wooden spoon. Following the script of a guided cooking show, Weinstein's accompanying video walks the home chef through the multiday process of making HFDC at home. With this recipe, the home and the factory merge, with 'DIY' coming to mean 'do industry yourself'.

Like reverse-engineered recipes for Coca-Cola, the HFCS kit opens a black box and promotes public amateurism. In her essay 'Reaching the limit: when art becomes science', Beatriz da Costa celebrates public amateurism as a way of accessing the sciences obliquely: 'Rather than attempting to achieve expert status within the sciences, artists have ventured to

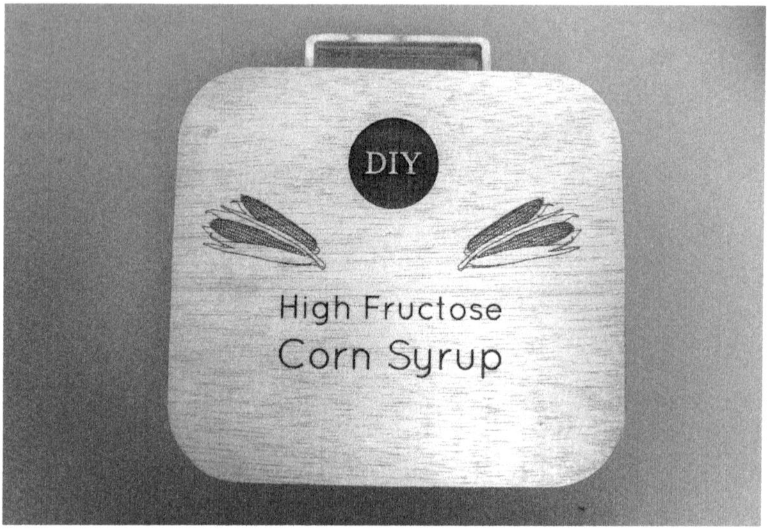

Figure 5.2 Maya Weinstein, *DIY High Fructose Corn Syrup* kit (2013). Courtesy of Maya Weinstein.

find help in the realm of hobbyism and do-it-yourself home recipes for conducting scientific experiments'.[11] Building on da Costa's definition of public amateurism, Weinstein promotes DIY HFCS as 'citizen food science': citizen food scientists uncover the invisible processes that animate industrial culinary technoscience. In the hands of the public amateur, the HFCS kit becomes the first step towards a more nuanced set of research questions about how we might alter or repurpose HFCS or how we might work with industrial corn varieties in unexpected ways. When discussed as public amateurism, making your own Coke or HFCS becomes a practice of dissent and a framework for questioning industrial food production. The do-it-yourself kit recalls our earliest encounters with science at home – the chemistry set (increasingly being replaced by the biotechnology set). The kit format places this project in familiar territory: although children and their chemistry sets do not necessarily grow up to be dissenting artists, chemistry sets do prepare us for the many recipes and instructions we will encounter at home, at work and at school, from cake mixes to laboratory protocols.

Like recipes, kits include ingredients, actions and a logical progression from one thing to the next. Of biotech kits, da Costa observes that

> similar to the pleasures of cooking one's dinner or fixing up one's house, the act of making, building and simultaneously learning seems to be an appealing and desirable way of spending spare time for those who can afford doing so.[12]

Kits connote leisure, discovery and curiosity. The kit format allows users to make something new using specialised materials and to inhabit a specific identity – chemist, biologist or even, in the case of da Costa's *Anti Cancer Survival Kit*, survivor.[13] Natalie Jeremijenko and Heath Bunting's *Biotech Hobbyist Magazine* produced some of the first biotech kits to circulate in an arts context. The magazine 'refolds social space by encouraging members of the public to become experimenters', using the language of kits and recipes to transform readers into active participants, enacting Higgins's 'thrust toward experience'.[14] The *Biotech Hobbyist Micropropagation Kit* includes everything you need to clone plants using plant tissue culture protocols. The *Starter Skin Kit* declares that 'it is time to wrest tissue culture from the privileged hegemony of the lab and relocate it to your kitchen', and invites users to cultivate tissues at home, with future projects elaborating on starter skin to create glowing and silicon-enriched versions of the initial culture.[15] Eduardo Kac's *Cypher* (2009) 'merges sculpture, artist's book and a DIY transgenic kit' to transform bacteria with DNA sequenced to reproduce the text of a poem.[16] The Tissue Culture and Art Project have made kits as well: *DIY DVK 01* (DIY De-Victimizer Kit, 2006) addresses feelings of guilt when consuming animals by keeping the tissues of those animals alive 'at least until the guilt recedes'.[17] As 'portable scores', kits invite rearrangement and innovation.[18] Like *Romances of the Meal* and *DIY HFCS*, these examples of kits and recipes in bioart contexts promote amateur investigations of new technologies. Once we can see inside the black box of industrial food (or industrial pharmaceuticals or industrial agriculture), new possibilities emerge for intervening in or resisting those systems.

Contemporary artists working with kits and recipes recall the *Fluxkit* and the readymade, as well as the long history of boxed collages and interactive kinetic displays. FLUXUS boxes and kits have been described as 'portable scores in the form of objects' – recipes of a sort.[19] Although Weinstein's kit functions in a more scripted and logical way than many of the poetic and even absurd activities proposed by Fluxkits, her work promotes a similar context collapse, asking participants to critically create. FLUXUS artists positioned their open-ended projects as activities akin to experiments performed in a laboratory; Hannah Higgins describes the 'thrust toward experience' of the Fluxkit.[20] Far from being prescriptive or rigid, Fluxkits are meant to be rearranged and invite unpredictable outcomes.

Bioart favours kits and recipes, inheriting the instabilities and possibilities of these forms from FLUXUS practice. Kits and recipes encourage the movement of protocols and substances from laboratory to kitchen and back again, remaking and connecting professional and amateur spaces.[21] Kits capture laboratory tools and techniques and allow for the dissemination of unusual ingredients. Kits are familiar, evoking DIY projects from childhood like chemistry sets, sea monkeys and tinker toys. Housing something unusual and radical in the recognisable format of the DIY kit facilitates the engagement and unpredictability cultivated by the makers of Fluxkits. Kits and recipes link industrial kitchens and citizen food science to growth, vitality and 'aliveness', qualities that have come to define bioart, while also promoting public amateurism. In the hands of bioartists, citizen food scientists and recipe artists, kits create activities and subject positions that allow for inclusive understandings of how food preparation might be interpreted as wet work.

6

Meat Culture

FBS (fetal bovine serum)

Thaw purchased fetal bovine serum (shipped on dry ice and kept frozen at −20°C until needed). Store 3 to 4 weeks at 4°C. If FBS is not to be used within this time, aseptically divide into smaller aliquots and refreeze until used. Store ≤1 year at −20°C.

Repeated thawing and refreezing should be avoided as it may cause denaturation of the serum.

In some cases, heat inactivation may be warranted. To inactivate FBS, heat 30 to 60 min in a 56°C water bath. Alternatively, FBS may be inactivated through radiation treatment.[1]

~

This chapter wrestles with collaboration and complicity by delving into how spaces of production interface with new food technologies. Reworking the use of 'embed' first coined by eighteenth-century naturalists, I argue that embedded bioartists use their status to bring lessons from the critical reinvention of Home Economics to biotech benches, often mirroring or referencing the kitchen laboratories that enclose us all. This story begins with my time in a repurposed biotechnology laboratory at the University of California Irvine. My participation in a workshop for artists on basic genetic engineering techniques illustrates how art laboratories, wandering biological materials and collaborative research practices combine to form unexpected relationships and alliances. SymbioticA, the first biotechnology laboratory dedicated to interdisciplinary collaborations between artists and scientists, conducts workshops all over the world that introduce artists, scientists and scholars to the genetic engineering of plant and animal tissues. At the 2005 University of California Irvine workshop, Oron Catts, artistic director of SymbioticA, and Gary Cass, researcher and technician, introduced basic materials and protocols in a laboratory context, while simultaneously presenting projects happening at SymbioticA and the histories of the various fields in which we were engaging, including tissue culture, agricultural science and molecular biology. The workshop was five days long, with each day presenting new information and providing opportunities to engage with biotechnologies in embedded, 'at risk' ways.

One of our days in the laboratory introduced us to a simple procedure to isolate DNA and a more complex exercise to create transgenic bacteria by introducing green fluorescent protein (GFP) to E. coli, producing a collection of glowing green Petri dishes. As described later in Chapter 9, 'Carnal Light', GFP is derived from bioluminescent jellyfish, so our transgenic E. coli contributed to the genetic drift of jellyfish biomaterials through a variety of biological contexts. After the GFP transformation, we were responsible for a living bacteria hybrid that glows in its Petri dish when poked with light, while its not-glowing relatives thrive in our lower intestines, and the jellyfish source of its glow drifts in the ocean.

Before beginning, Cass and Catts asked the group if we had any reservations about our participation in these practices. They introduced the day's work with a statement and a question: 'You are about to be implicated in genetic engineering. Are you sure you want to go on?'[2] This question was largely rhetorical. We were all there, after all, for a workshop which we intended to complete, and by eating genetically modified foods and using vaccines, insulin and other pharmaceuticals we were already caught in the nets of genetic technologies, although most of us had not wilfully produced a genetic modification before that day.

As a group, we agreed to accept responsibility for the engineered tissues we created during the workshop. Following the material results of our work beyond their immediate practical applications had not been part of my prior encounters with science laboratories. In previous biology classrooms, responsibility for materials was framed in terms of cleaning up. Being tidy trumped more serious considerations of life, death and moral obligation. Even when living and dead things were part of our experience, these creatures just went away somewhere after class. I remember hatching chicks from eggs under warm incubator lights and dipping my hands in plastic bucket coffins, filled with formaldehyde, to pet floating dead frogs. What happened to these frogs and chickens? Were they donated to a local farm, flushed down the toilet, tossed in the dumpster? Although ultimately UC Irvine workshop participants were unable to visit the facility where biohazard materials are incinerated, I appreciated that we ought to round up everything we produced and personally deliver it to its crematorium death. It was only right to accompany our Petri dishes to their small biohazardous graves.

Artists employ 'wetware' to encourage this type of responsible care for the living things nurtured by labs, human and non-human alike. Whether that means imagining scenarios for creating and interacting with fluorescent plants and animals or freely sharing genetic modifications to subvert connections between biotech and capitalist economies, conversations between artists and scientists yield new ways of doing and thinking about science. My experience of becoming responsible to transgenic organisms at the SymbioticA workshop reflects the goals many bioartists have for the reception of projects that manipulate living plant and animal tissues as media. Bioart introduces audiences to situations that demand ethical engagement and care.

At UC Irvine, I felt a recursive effect between the cultures held in our hands and the laboratory holding us within its walls. As visitors to laboratory spaces, we potentially affected those spaces with our ongoing engagement and dialogue. In the laboratory – from *laborare*, 'to labour' – work happens. This simple definition would appear to invite radically interdisciplinary labour but until recently, laboratory spaces have not been the province of artists. The result of art–science convergences related to (but not wholly attributable to) advances in biotechnologies and genomics, increasing numbers of what I call 'art laboratories' encourage hybrid work environments. Art laboratories immerse participants in science practices by way of critical engagement with 'technologies of living substance'.[3] The art laboratory becomes more than a work environment or a stage for displaying predetermined cultural or scientific knowledge. Art laboratories emphasise encounter, exchange and participation, enacting the contingencies and interdependencies that often remain beneath the surface of analyses produced by more easily categorised and disciplines contexts.

Laboratories are defined by their occupants, experiments and equipment but are also defined by their location and the actions that are possible there. Laboratories constructed for artistic production have different priorities than laboratories made for scientific research or corporate research and development. According to Christopher Henke and Thomas Gieryn, laboratories hold contradictory states in tension, seeking to stabilise, through architecture and place, 'public and private, visible and invisible, standardised and differentiated'.[4] Art laboratories address these tensions by employing the rhetoric of public experimentation, so the status of public experiments in art and culture becomes important to their definition. This includes the idea that 'acting like' or 'acting as' a scientist or being 'embedded' within a laboratory produces cultural meaning, and the possibility that direct encounters with scientific protocols activate concerns about privacy and visibility.[5]

Unlike the amateur, the embedded collaborator works alongside experts in institutionally sanctioned ways, whether as a research fellow or a director of a cross-disciplinary laboratory, roles Oron Catts has taken up at various points in his career as a 'tissue engineering artist'. As the Tissue Culture and Art Project, Catts and Ionat Zurr were research fellows at the Tissue Engineering and Organ Fabrication Laboratory at Harvard Medical School, and then founded SymbioticA at the University of Western Australia. Artists devoted to science/art collaboration are embedded because they integrate with spaces and entire disciplines in an effort to merge their practices with those of scientists and technicians. Of course, it is possible for the same practitioner to occupy both positions at various points in their career. Returning to Catts and Cass's SymbioticA workshops, the travelling workshop emphasises public amateurism and non-institutionally sanctioned modes as identities for participants to try on, but the SymbioticA laboratory itself is institutionally sanctioned, as was the Tissue Culture and Art Project's Harvard residency.

Nicola Triscott articulates these different modes as spatial relationships: 'A growing number of artists have started working within scientific contexts [...] [while] other artists have chosen to work outside the structures of science, appropriating science as amateurs or tactically exposing the systems that fund and promote science.'[6] Bioartists focused on food occupy the breadth of 'inside/outside' locations Triscott describes, but more often than not exist in a grey area between inside and outside. A closer look at these different ways of engaging the laboratory reveals the ways in which bioart crosses spatial and disciplinary zones in its collaborative production. The laboratory as a fictional and real site of future food and 'Frankenfood' animates many of these projects, which build on the legacies of interdisciplinary feminist art to reveal the degree to which such future food scenarios have already arrived.

To be embedded implies both permission and proximity. First used to describe rocks, fossils and mountains, to 'embed' something is to surround or fix something in another material. A March 2007 draft addition to the Oxford English Dictionary defines 'embed' as 'to attach (a journalist) to a military unit to report on a conflict'.[7] This newest meaning of embed, originating in United States military jargon, has been employed to describe art and science interactions. With its origin in a context of global war, this is the only use of the word that suggests people interacting with other people. Comparing bioartists to journalists among soldiers implies that art has explanatory, illuminating and interpretative functions when produced from the inside, and also implies that scientists are aligned with soldiering, which may be true when considering funding sources for the sometimes-unexpected science research that has military applications.

The figure of the journalist/artist in a war/laboratory context helps me work through how artists make processes public, and how art shapes the terms of public debate. Kylie Tuosto suggests that 'the effect of embedded reporting on the American public is a distraction from and desensitisation to war, as well as a perpetuation of American overconfidence in military ability'.[8] This analysis of state policy and how information becomes public evokes the logic behind critiques of bioartists as puppets of corporate interests: how can an 'embed' be critical of the environment that structurally supports their work? Using 'embed' as pronoun, name and synonym for 'artist collaborator' contextualises the military funding that flows through many universities and science laboratories and the empire-building, colonial function of that funding and those labs.

'Embedded' also signals ecological interdependence, a nested organism. In the case of *in vitro* meat, the Tissue Culture and Art Project's *Disembodied Cuisine* shaped the popular and scientific terms of public debates about laboratory-grown meat.

The Tissue Culture and Art Project has two members at present, Oron Catts and Ionat Zurr. Catts calls himself a 'tissue culture artist', while Zurr prefers 'wet biology art practitioner'.[9] Catts and Zurr are committed to an art practice that incorporates biological protocols, with tissue engineering as a primary focus. Catts practices what he calls 'pleasurable tissue culture', emphasising the sensual and aesthetic enjoyment of tissue culture processes, which are often beautiful to perform and observe both under the microscope and with the naked eye. Guy Ben-Ary was a member of the Tissue Culture and Art Project from 1999–2003 and collaborated with Catts and Zurr on *Disembodied Cuisine*, and the Tissue Culture and Art Project also works closely with researchers at their host organisation, SymbioticA Centre of Excellence in Biological Arts at the University of Western Australia, where Catts is Artistic Director.[10] Since its inception, the laboratory has garnered more support from the university, and now contributes to university curriculum. In addition to offering undergraduate and postgraduate classes and awarding degrees such as the Master of Biological Arts, SymbioticA members travel to conduct workshops around the world like the workshop I attended at the University of California Irvine.[11] While giving an overview of their history with tissue culture as an art medium, Zurr and Catts write:

> when the manipulation of life takes place in an atmosphere of conflict and profit-driven competition, the long-term results might be disquieting. One role that art can play is to suggest scenarios of 'worlds under construction' and subvert technologies for the purpose of creating contestable objects.[12]

The sculptures, performances and spaces Catts and Zurr have produced over the last 15 years have grown and changed by extending species boundaries and interrogating zones of human/nonhuman encounter like food and eating. Their 2003 installation, *Disembodied Cuisine*, presents amphibian meat that was conceived of and prototyped in a hospital for humans.

The Tissue Engineering and Organ Fabrication Laboratory at Harvard University/Massachusetts General Hospital designs and builds 'living replacements for organs and tissues' in collaboration with MIT's Draper Labs and the Center for Regenerative Medicine.[13] Catts and Zurr were residents in the laboratory in 2000–1, where they developed projects they had been working on since their earliest collaborations with Professors Miranda D. Grounds, Traian Chirila and Dr Stuart Bunt at the University of Western Australia's School of Anatomy and Human Biology. The Laboratory for Tissue Engineering and Organ Fabrication website describes the Tissue Culture and Art Project residency:

> Oron Catts and Ionat Zurr have received a grant, from the New Media Arts Fund of the Australia Council, for a one-year residency in the lab [...] Tissue engineering holds much promise for improving the quality of human life. However, tissue engineering for artistic purposes has largely been overlooked. We are growing tissue sculptures, 'semi-living' objects, by culturing cells on artificial scaffolds in bioreactors. Ultimately, the goal of this work is to culture and sustain, for long periods, tissue constructs of varying geometrical complexity and size, and by that creating a new artistic palette.[14]

The laboratory lists 'Art' alongside research areas like 'Kidney', 'Liver' and 'Bone & Cartilage', bringing artistic production into dialogue with research on the body's mechanics in a hospital setting. The Tissue Culture and Art Project had close working relationships with scientists before their residency at Massachusetts General, but this description of their work conveys the collaborative, embedded methodology taken up by artists who choose to physically locate their practice inside institutionally-sanctioned laboratory spaces, while the vocabulary of the 'new artistic palette' translates how tissue culture art might unsettle more familiar modes of art making. The living palette of tissue culture art does not represent or mirror the world but rather fabulates and grafts onto the world. This palette does not precipitate an act of painting, sculpture or image capture but instead unfolds as many small, slow, fragile, slimy becomings that require care from their makers.

Human/nonhuman categories are already troubled in tissue culture practices, even without expanding such practices to include the category of 'art'. In her essay 'Living differently in time: plasticity, temporality, and cellular biotechnologies', Hannah Landecker asks tissue culturists how they worked and thought differently when human versus non-human or plant tissues were circulating in their laboratories. She found that tissue culture specialists 'were, to a person, absolutely disinterested in my questions about distinctions between human and animal matter in their laboratories, responding with much more detail and emotion on quite separate issues of practice'.[15] Landecker's project turns away from the category of the human and towards techniques and questions of cultural significance and investment that have also become central to the Tissue Culture and Art Project's work. She and the Tissue Culture and Art Project both consider how approaches to the nonhuman ultimately speak to how humans care for ourselves as a species: 'Most bluntly, how we handle nematode matter or yeast matter or chicken matter may be more formative for what we do with and how think about human matter than any particularity of human matter as human.'[16] The Tissue Culture and Art Project shares this point of view with the tissue culturists Landecker interviewed: human particularity is not an essential quality but an affinity that may be tried on or stripped away, revealing and concealing human and animal bonds in uncanny ways.

As new processes and methodologies opened up to them, like the scientists Landecker interviewed, the Tissue Culture and Art Project 'sloughed the human',[17] focusing less on human narratives and artefacts and more on how tissue culture protocols might complicate human relationships with nonhuman others. As art and science researchers deeply inhabit the 'tissue culture point of view', species boundaries and the primacy of human qualities become less pivotal. Catts and Zurr write: 'when we started to look at what we could get these tissue constructs *to do*, we emulated humans' path of intersection with fellow living beings'.[18] Pathways matter more than definitions or characteristics of actors in the tissue culture network, and eating-as-exploitation immediately follows this understanding, making the lives of semi-livings, even semi-livings made from human tissues, akin to the lives of animals. Monika Bakke describes this emphasis on 'fellow living' and excitation about involvement with living matter as '*zoe*-philia', the transgressive desire and preference for base life over the life of the individual or (human) community:

> Contemporary *zoe*-philia [...] differs significantly from any of its previous versions due to the impact of biotechnological advancements and our awareness – informed by molecular and cellular biology – of whom we are in relation to other lifeforms [...] *Zoe*-philic desires influence contemporary debates about life itself, human exceptionalism and human-animal relations as much as do biotech industries.[19]

Bakke points out that the Tissue Culture and Art Project does not work on 'the level of representation, but on the level of actual intervention into living systems', what Robert Mitchell would call the 'vitalist' approach to bioart.[20] This direct engagement allows for *zoe*-philic approaches to life. Bakke describes the pathways Tissue Culture and Art Project explores as part of a 'shift in our relation to life itself' – for the Tissue Culture and Art Project, '*zoe* (life) [is] a trans-species generative force'. *Zoe*-philia explodes the boundaries of organisms, risking a return to the outdated language of instrumental immortality currently used by transhumanists but also inviting the transspecies entanglement and responsibility advocated by cross-species theorists like Donna Haraway. The '-philia' suffix also suggests that by falling into *zoe*-philia we fall into a love that can never be reciprocated, as our object of love is so radically different, even fetishised.

Disembodied Cuisine presents an opportunity to fall into *zoe*-philia and to examine human–animal relationships through the familiar interface of food and eating. As installed at 'L'art biotech' in Nantes, *Disembodied Cuisine*'s dark laboratory space consists of a black dome punctuated by round portholes through which visitors can peer into the tissue culture laboratory.[21] From another vantage point, the sterile hood becomes visible in its entirety, revealing the artists at work with the tissues. In contrast to the dark, even ominous laboratory, the light and spare dining room floats inside clear plastic sheeting printed with biohazard symbols. A small group of frogs overlook the dining space from their well-appointed aquarium home. For *Disembodied Cuisine*, the Tissue Culture and Art Project employed a cell line that originated from the live, aquarium-dwelling amphibians and then seeded their cells in polymer scaffolds. These cells multiplied into very small portions of cultured frog steak. The tissues were housed in an on-site bioreactor during the exhibition, and at the end of the growth period, invited guests sampled the small steaks after they were prepared with a garlic and

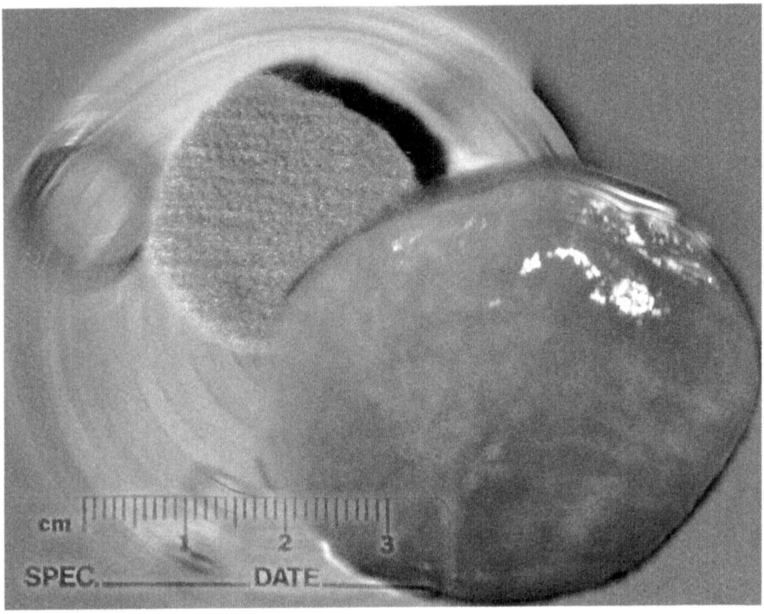

Figure 6.1 The Tissue Culture and Art Project, *Tissue Engineered Steak No.1* (2000), a study for *Disembodied Cuisine*. Pre-Natal sheep skeletal muscle and degradable PGA polymer scaffold. This was the first attempt to use tissue engineering for meat production without the need to slaughter animals. Part of Oron Catts and Ionat Zurr's Research Fellowship in the Tissue Engineering and Organ Fabrication, MGH, Harvard Medical School. Courtesy of the Tissue Culture and Art Project.

parsley sauce. The installation and 'feast' explore the idea that meat might be grown from animals while the animals are still alive.

The first iteration of this project, produced under the auspices of the Tissue Engineering and Organ Fabrication Laboratory at Harvard University/Massachusetts General Hospital, used prenatal sheep cells.[22] In that prototype, the cells used to seed the polymer base were taken from an unborn animal, whereas in *Disembodied Cuisine* the cells used in the tissue culture process are from a cell line with representative donor organisms living in an aquarium overlooking the installation. This relationship is important: in the art exhibition context, unhatched cells are more abstract than the live frogs inhabiting an aquarium while their tissues expand *in vitro*. In a biology research context, the inverse could be true: the process of metamorphosis inspired some of the earliest explorations of tissue engineering.[23] Involving tissue culture art with food, eating and disgust sets this work apart from other semi-living projects, bringing the semi-living into familiar domestic and commercial spaces: the kitchen, the dining room and the restaurant. The growing tissues must be constantly fed with nutrient-rich fetal bovine serum (FBS), and are housed in purpose built bioreactors, which keep their temperature and oxygen supply constant. Visitors to the installation can participate by 'feeding' the cells, which in turn are fed to diners at an invitation-only feast, the culmination of the installation. The Tissue Culture and Art Project proposes art as both food and 'semi-living' food that needs to be fed.[24] By locating these feedings in domestic dining spaces, the Tissue Culture and Art Project works with food and eating and with the social space of the dining room, the dining table and even the aquarium. In this meal more than most, the aquarium brings diners into a relationship of responsibility with its inhabitants.

By the time the Tissue Culture and Art Project produced *Disembodied Cuisine*, the group had established that 'taking the living semi-living entities outside of the scientific laboratory into a lab in a public space is not trivial but of enormous importance to us'.[25] As with the work of Critical Art Ensemble, the public performance of science allows the Tissue Culture and Art Project to make connections between materials and ideas that would not be possible, or at least would not be as visceral and engaging, if documentation of laboratory work was shown in place of the work itself. *Disembodied Cuisine* suggests that *in vitro*, raw and rare meat must be considered a 'semi-living' entity, and that meat has a synechdochic relationship with the healing frog in the aquarium. Guests eat the frog's wound. A typical steak dinner does not add up to a 'semi-living' entity, but conceptually, rare meat engages similar issues: meat should be alive enough to be recognisably animal-based, but not so alive that it moves or fights back. A steak is both part and outside of a cow. Vast agricultural systems nourish and destroy animals destined for meat production, and most people consider the death of an animal to be a more controversial event than the death of a plant or a tree. Cells within cuts of meat (especially if the bones are left whole) stay alive for hours or even days after the animal is slaughtered. These cells, no longer part of a living organism, could be considered 'semi-living' in that they continue their half-lives independent of a larger, intelligent whole.

As I follow *in vitro* meat through its many forms and contexts, the conjoined activities of eating and killing become central concerns when considering how semi-living entities complicate our understanding of both. Cultured cells often outlive their organism, making artificial meat a human–animal interaction open to long-term engagement that surpasses the span of human lives. Catts and Zurr have named the physical manifestations of this extended life the 'extended body'. Killing shifts in scale, with the possibility of micro-level kills happening on a vast scale, killing and eating the same tissue made from the same cells, over and over again – a logic of reproduction founded on the coupling of prolific growth with countless tiny deaths from incidental and intentional contamination. Donna Haraway writes: 'outside of Eden, eating means also killing, directly or indirectly, and killing well is an obligation akin to eating well'.[26] The inevitable exploitation of the semi-living is a constant theme in the Tissue Culture and Art Project's work: how are we responsible for the semi-lives and the small, repetitive deaths of the artificial tissues we create for organ research, food, and art? Are tiny deaths tiny killings? And how might we kill well in miniature? The Tissue Culture and Art Project considers eating to be 'the ultimate exploitation [...] we observe that this form of relationship is the most problematic but also the most primal, and brings us back to the basic interaction humans have with their fellow living beings'.[27] We have built industries and architectures around this 'ultimate exploitation', with social systems, menus and factory farming devoted to these 'basic interactions'. With its simple split floorplan, *Disembodied Cuisine*'s cavernous laboratory becomes the shadow cast by the bright dining space, implying that behind every dining room lurks a laboratory, a shroud and a fecund killing floor.

Extended bodies are made up of thousands of pounds of cultured tissue alive *in vitro* around the world. How can one discrete physical entity yield infinite iterations of itself yet not-itself? The frog steaks produced in *Disembodied Cuisine* are just a fraction of the frog's larger extended body; while caring for, eating, or killing these cells, other cells living elsewhere propagate to produce 'a unified body [of] disembodied living fragments'.[28] The frog's body in fragments troubles our ideas about corporeal boundaries; one frog exists simultaneously in multiple places. The Tissue Culture and Art Project presents this sense of disconnect vividly with *Disembodied*

Cuisine; less obvious are the vastness and material excess of thousands of pounds of 'semi-living' *in vitro* biomass dispersed across innumerable laboratories and research facilities.

Artificial meat research happens in conversation with research into artificial organs and muscles. In order to taste better than the frog steaks in *Disembodied Cuisine*, *in vitro* meat will have to move, either internally via a system of capillaries and electric muscle pulses, or externally via physical stretching and manipulation. Although I did not taste the frog steaks, from the video and photo remains of the project and the comments of people who did eat the frog steaks, the problem was the texture of the meat, which was similar to gelatine. Furthermore, the polymer fibre base did not dissolve (I have handled the fibres and would imagine that chewing them would be similar to chewing felt or moleskin). Even if the polymer bases had dissolved as anticipated, the meat would have had the jiggling, formless qualities of processed, waterlogged gelatine. While it is tempting to declare that *in vitro* meat should be pushed to taste and feel like more familiar forms of meat, gastronomists like Hervé This and chef-engineers like Ferran Adrià might favour the opposite approach if presented with the ingredients of *Disembodied Cuisine*. Perhaps it is the human palate that should be pushed to appreciate the uncanny difference of *in vitro* meat. To some degree, this has already been accomplished with plant-derived products like tofu dogs and fakin-bacon, which have been in circulation long enough that consumers have grown accustomed to regarding smooth extrusions of textured vegetable protein as meaty. *Disembodied Cuisine* offers a radical departure from the tofu dog: replacing chewiness with a quivering gel, these frog steaks offer a taste of the plasticity of polymer scaffolding technologies. This is a medium-specific meal with haptic and proprioceptive qualities appropriate to the technologies that produced the food.

The Tissue Culture and Art Project does not privilege immortality, preferring to focus on ethical choices in a mortal, fleshy world, not a future world of glass hearts and supermen. This approach owes much to 'ecocritique' and ecofeminist interventions.[29] *Disembodied Cuisine* emphasises the triangulated relationship between human eating bodies, the living frog in the aquarium and the frog's extended body on the dinner plate and in laboratories around the world. All of these complications, market forces, economies and systems of exploitation do deeper work toward sorting out our post-human interaction with animals as food and as partners in the possibility of *in vitro* vegetarianism. Rather than revelling in the excitement of the extended body simply existing or, as with Carrel's chicken heart, the possibility that the frog's extended body has outlived its mortal form, each frog steak evokes relationships between its producers, its consumers, humans, scientists who use frog cell lines, laboratory frogs, wild frogs and farmed frogs. All of these relationships commingle in each bite of frog steak. Writing about the ethics and futures of *in vitro* meat products, Jacob Metcalf suggests that 'at its best cultured meat promises to be sustainable merely by *not taking away* because it appears able to stand outside of ecological relations'.[30] *Disembodied Cuisine* engages this critique by firmly situating *in vitro* meat within a web of social and ecological relations. Domestic and laboratory spaces are opened up to emphasise the many ways that *in vitro* meat fails to promote sustainable relationships within social and biological human–animal ecologies.

By making and eating *in vitro* meat, *Disembodied Cuisine* brings a practical, fleshy approach to meat consumption, a topic that has been on the research agendas of animal rights activists and tissue culturists for a very long time.[31] Eating, whether one's diet is meat or plant based, will always be 'one of the most common zones of interaction between humans and other living systems'.[32] *Disembodied Cuisine* entered cultured meat research as a pioneering practical application of a scaffolding technique that has been imagined by others in theory.[33] The organisation New Harvest ('working to develop new meat substitutes, including cultured meat') cites SymbioticA as 'the first laboratory that tried to produce meat *in vitro*'.[34] An article (written by several New Harvest affiliated researchers) in the journal *Tissue Engineering* explains the scaffold technique and cites *Disembodied Cuisine* as the first successful instance of scaffold-based *in-vitro* meat production:

> In scaffold-based techniques, embryonic myoblasts or adult skeletal muscle satellite cells are proliferated, attached to a scaffold or carrier [...] and then perfused with a culture medium in a stationary or rotating bioreactor [...] These cells fuse into myotubes, which can then differentiate into myofibers. The resulting myofibers may then be harvested, cooked, and

> consumed as meat. van Eelen, van Koooten, and Westerhof hold a Dutch patent for this general approach to producing cultured meat. However, Catts and Zurr appear to have been the first to have actually produced meat by this method.[35]

Here, art/science collaboration sets a precedent for scientists researching the same field of inquiry. The article's authors do not feel compelled to qualify SymbioticA research as art. 'Catts and Zurr' are referred to in the same manner as 'Benjaminson, Gilchriest, and Lorenz', or any other researchers in pursuit of viable cultured meat. This distinction recalls Robert Mitchell's distinction between 'prophylactic' and 'vitalist' modes in bioart:

> Whereas the prophylactic tactic is premised on the principle that art best engages the problematic of biotechnology by re-presenting it in other media, the vitalist tactic, by contrast, is premised on the principle that art best engages the problematic of biotechnology when it becomes itself a medium for this latter.[36]

Disembodied Cuisine seeks to make present, rather than represent, tissue culture as culinary technoscience.

The scientific and historical richness of the Tissue Culture and Art Project's work depend on frogs and frog lives. The frogs of *Disembodied Cuisine* are of two varieties: *Xenopus laevis*, the toad source of the manufactured cell line used to create the frog steak cultures, and *Rana kl. esculenta*, frogs often farmed for food in France. Catts and Zurr describe the aquarium environment: 'frogs could frolic amongst miniature Venus sculptures and ultimately observe the ceremonial supper before finally being released unharmed in the nearby botanical garden [...] The Venus figures in the aquarium raise the question of the possibility of non-anthropocentric art.'[37] Like the bodies of the Venus statuettes, the aquarium's vegetation marks these frogs as animals on display. They live in a small garden, enclosed in a bigger garden. These display animals, framed in a carefully constructed mise-en-scène, might even be said to possess a measure of Laura Mulvey's 'to-be-looked-at-ness'.[38] The Venus statues force viewers to reflect on the anthropocentric gaze. How do our eyes move differently over feminine art bodies versus frog bodies? Taking

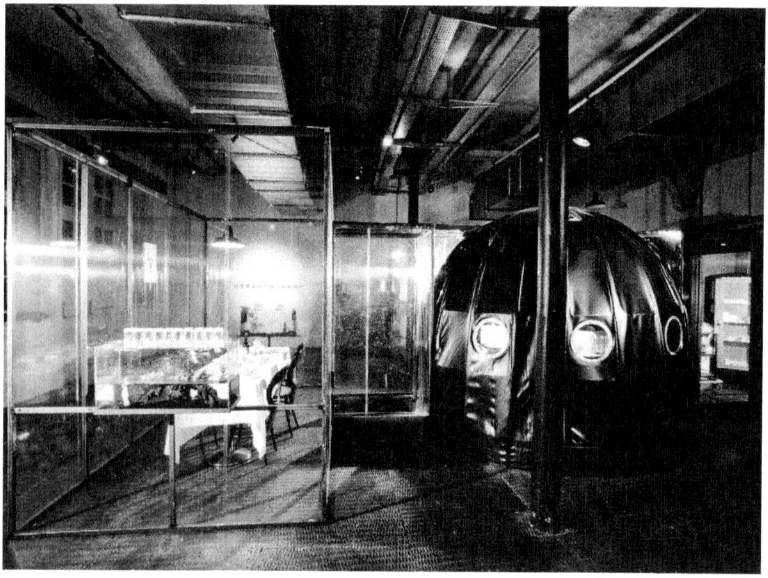

Figure 6.2 The Tissue Culture and Art Project, *Disembodied Cuisine* installation, Nantes, France (2003). Photo: Axel Heise.

a decorative garden space as the field location for frogs extends questions of looking and spectatorship by embedding frogs in visual structures different from those mapped onto landscapes that supply laboratory frog populations and different from landscapes created for frogs on public display. In Nantes, frogs may have been raised in labs, on farms, or plucked from ponds polluted by pesticides, as with Rachel Carson's frogs in *Silent Spring*.[39] After the installation, the frogs were released into the local botanical gardens, a setting that productively muddles all of these possible outcomes for frog lives. As aesthetic and educational spaces for safe visual encounters with familiar and exotic plants, botanical gardens cultivate certain animal populations and discourage others, often with homemaking interests in mind. The botanical garden mirrors the real or the imagined garden of the visitor.

That frogs are cultured in order to effect material changes at the level of taste and haptic oral encounter demonstrates how much tissue culture has changed in our popular imagination, that is to say, how much the 'tissue culture point of view' has shifted. The Tissue Culture and Art Project chose to work with frog cells in part because of the installation's location in France, one of only a few places in the world where frog meat is popular for dinner:

> We titled the installation *Disembodied Cuisine*, playing on the notion of different cultural perceptions of what is edible and what is foul. We grew semi-living frog steaks, with the intention of raising questions about the French resentment towards engineered food and the objection by other cultures to the consumption of frogs.[40]

Eating signals interdependence, and the eating body's relationship with frogs points to a telling collapse between research objectives and culinary objectives. The Tissue Culture and Art Project intended to embed the installation in several different debates at the same time, playing French objections to bioengineered food against their embracing of frog meat for dinner.

Frog lives connect this installation to the earliest successful tissues cultured in laboratories. Alexis Carrel biographer Theodore Malinin notes that 'the successful cultivation of amphibian nerve cells by the American researcher Ross G. Harrison in 1907 attracted wide attention among his contemporaries', both because his experiment ended a debate about how nerve cells grow and because Harrison developed new tissue culture techniques.[41] Carrel's assistant Montrose Burrows visited Harrison's laboratory and adapted his technique to warm-blooded animals. Amphibians were also important to early perfusion attempts. Carrel describes the first perfusion apparatus, tested in 1868: 'In this manner, de Cyon kept the heart of a frog beating for forty-eight hours.'[42] Frogs are everywhere in early tissue culture experiments, making the Tissue Culture and Art Project's decision to culture frog tissue a recontextualisation of the first tissue culture experiments ever performed. In her book *Crystals, Fabrics, and Fields: Metaphors That Shape Embryos*, Donna Haraway highlights Harrison's concern with the material semiotics of the artist: 'Ross Harrison did not write directly about aesthetic issues, but his entire life's effort, in its content and its form, betrayed his adherence to dictates similar to the artist's.'[43] These 'dictates similar to the artist's' animate a career that spans decades of unrest and uncertainty about form and methodology in biology (art of the early and mid-twentieth century was similarly preoccupied by process and method). Haraway's description of Harrison's oeuvre shows how the relational, process-driven work of tissue culture effects and contains change. Shifts in the tissue culture point of view from Harrison's era to the Tissue Culture and Art Project's can be roughly described in terms of a movement away from extension, mapping and, for Carrel and Lindbergh, 'immortality', and toward discourses of material semiotic world-changing: rather than living forever, today's tissue culture point of view emphasises living interdependently.

Entanglement, interdependence and the recognition of ecological consequences of factory farming inform current *in vitro* meat research, which emphasises two areas: developing a 'vegetarian' growth medium system, and exercising cultured meat to improve its texture. Most proponents of cultured meat position tissue engineering technology as an environment and animal-friendly alternative to regular meat, implying that *in vitro* meat respects animals and the earth. This isn't necessarily the case. In her effort to make these 'abstract' animal products materially present to organisations like PETA (People for the Ethical Treatment of Animals) and New Harvest, Ionat Zurr demonstrates how new technologies do not promote progressive ways of engaging animals and their suffering:

> The abstraction of these animal products in the technology associated with tissue culture served to obscure the very real victims from the eyes of organisations such as PETA and the European Coalition to End Animal Experimentation. For example, on a rough estimate based on TC&A experience in growing in-vitro meat, growing around 10 grams of tissue will require serum from a whole calf (500ml), which is killed solely for the purpose of producing the serum. Furthermore, at the request of a researcher (who later founded New Harvest, a spin-off company of the University of Maryland that is a nonprofit research organization working to develop new meat substitutes, including cultured in-vitro meat) we were asked to calculate the costs of such meat. We calculated that one gram of in-vitro frog steak we grew in Nantes in 2003 cost us then US$650 (this is without calculating in monetary value the costs for the environment such as discarded plastic ware, energy etc).[44]

Zurr's analysis performs the embedded position the Tissue Culture and Art Project have adopted in their work. Their work in and creation of laboratories allows the Tissue Culture and Art Project to make critical observations about standard practices and financial costs that would exist within science laboratories as budget items rather than research results.[45] *Disembodied Cuisine* focused on animal suffering by addressing our participation in systems that relocate and sterilise but do not eliminate the slaughterhouse.

Since 2003, laboratories around the world have been prioritising research into physical exercise for semi-living tissues and vegetarian growth media. As mentioned previously, the texture of cultured meat depends on aligning muscle fibres by exercising, either by means of a scaffold medium that expands and contracts or by a mechanical stretching method. Even for a product that is relatively mushy, like sausage filling or dog food, the difference between meat that has been exercised and meat that has not is the difference between hamburger and marmalade. This distinction was played out in *Disembodied Cuisine* when feast participants were unable to chew and swallow their frog steaks – the wobbly jelly structure coupled with the fibrous scaffold material proved too disturbing to the palette. Current scenarios envisioned for stretching meat include printing sheets of cells on a stretchy surface, assembling vascular systems to help construct more complex 'cuts' of *in vitro* meat and adjusting temperature and pH to produce small changes in the surface area of the scaffold.[46] Catts and Zurr have prototyped an exercise system for 'disembodied livestock' that builds on research into physical stretching and manipulation of tissues.[47] New Harvest proposes a glycerine-based medium; the aforementioned article in *Tissue Engineering* discusses alternate growth media already in development, including systems that employ maitake mushroom extract and lipids. One suggestion for how to improve tissue culture medium systems involves working with liver cells to generate the growth factors necessary for cell growth. Liver cells could form part of 'coculture systems' that would nourish *in vitro* meat without depending on fetal calf serum.[48]

Enter People for the Ethical Treatment of Animals, 'the largest animal rights organisation in the world'.[49] At first it seems obvious why PETA would be interested in *Disembodied Cuisine*: here are animals peacefully living out their lives while human beings eat meat created from their tissues. As told in Zurr and Catts's essay 'The ethical claims of Bioart: killing the other or self-cannibalism?', Ingrid Newkirk, PETA cofounder, offered her own tissue as a base for steak:

> Following Peter Singer's utilitarian ethics and the rejection of speciest perceptions, we were 'entertaining' ourselves with the idea of creating semi-living steak made out of an adult consenting human. PETA were offering us this human; their organization director Ingrid Newkirk. Furthermore, Newkirk suggested eating her own flesh.[50]

The Tissue Culture and Art Project decided against Newkirk's self-cannibalism strategy. Zurr and Catt's essay does not mention PETA's second point of involvement with the Tissue Culture and Art Project and the caveats placed on their financial participation. SymbioticA rejects funding from science industry groups and special interest groups like PETA. In a panel discussion at the UCI SymbioticA workshop, Catts speculated that PETA might never have offered support had they been aware of the animal-based growth medium system used in creating *Disembodied Cuisine*.[51] As an alternative to participating in SymbioticA's project, either financially or via cell donation, Zurr and Catts suggested that PETA support research on a vegetarian alternative to fetal bovine serum. PETA responded that they are a public relations organisation, not a research organisation.[52] This response betrays a correlation between art and PR that denies the research value of art projects. Why is creating a steak from Ingrid Newkirk or a frog less of a research activity than discovering non-violent alternatives to fetal bovine or calf serums? Does one have more public relations value than the other? Perhaps PR worlds find the 'art' element in the steak project to be more seductive than the 'research' element in tissue culture serum technology. This reading of tissue culture art practices as research proposes a relationship with art that does not privilege visual perception or aesthetic judgement. Instead, viewing art as research emphasises the intellectual and discursive participation and transformation of a variety of participants, including but not limited to the artists themselves.

Acquiring this research reputation is both the advantage and the challenge the artist faces when engaging with technology as an 'embed'.

Given that the meat industry has compelled us to engineer animals whose bodies express science fiction characteristics, one might wonder if projects like *Disembodied Cuisine* consciously or unconsciously reify industrial meat production and dystopian science fiction futures. In her essay 'Leonardo's choice: the ethics of artists working with genetic technologies', Carol Gigliotti wonders why meat is on the Tissue Culture and Art Project's table: 'Why continue to use animals in any form as food if you wish to question the traditional view of the non-human?'[53] Gigliotti goes on to argue that projects like *Disembodied Cuisine* are anthropocentric at heart:

> The continuation of art practices of creating life-forms through biotechnological means can only serve to implicate these practices, and artists who are involved in them, in contributing to a worldview that still values particular human needs above all else. This worldview, based as it is on the control and manipulation of nature, will continue to blind us to the more radical transformation of acknowledging that we have always been transgenic.[54]

Readings like this illustrate the double-edged position of the artist who chooses to be an embedded collaborator. For Gigliotti, the laboratory engulfs the embedded artist to such a degree that all critical actions are predetermined by the scientific discourses and work philosophies surrounding the artist. Gigliotti reads laboratory work and 'the control and manipulation of nature' as barriers blocking humans from recognising new forms of kinship that reject anthropocentric privilege.[55] Embedded artists are seen here as equivalent to laboratory scientists, but their activities are much more complex. Embedded artists perform discursive work that often closely resembles the 'lab study' of science and technology studies, but artists produce more than studies; they produce their own laboratories. *Disembodied Cuisine* reveals and participates in the laboratory slaughterhouse that manufactures fetal bovine serum. But by using the language of victimisation, harvest and healing to describe the installation's frogs (rhetoric artists can adopt with more ease than scientists), *Disembodied Cuisine* forecloses the anthropocentric privilege and pleasure of slaughter. Because the frog and the extended body of the frog are both alive, this absurd, transgressive meal of semi-living frog tissue offers a compelling argument for relating differently with animals in one of the most pervasive and normalised spaces of their exploitation – the dining room.

To conclude this chapter, I turn to Science and Technology Studies (STS) to continue asking difficult questions about complicity and participation when working inside laboratories of different kinds. Ionat Zurr recognises the risks of the embed's complicity as a form of 'suffering':

> Artists working with life manipulations, and more precisely with biotech, are participants in that culture. Besides the important act of democratising these technologies for the wider public, artists can 'suffer' from the embedded position; hence identification and participation with the situation they came to investigate and report on objectively [...] In the case of the critical artist, how does she resolve the paradox of using the technologies she is critiquing or working with in the context of engaging with an economy she is critiquing?[56]

The answer to this question resembles a lab study. In his essay 'Give me a laboratory and I will raise a discipline', Park Doing calls on future laboratory studies scholars to carry out the promise of the earliest ethnographies of laboratories and 'go directly for [...] a particular fact and wrestle with how its endurance obtains within the "*in situ*" world of practice'.[57] Bioart projects answer Zurr and Doing's calls to action by performing the contestability of scientific, artistic and industrial intersections. In the essay whose title Doing adapts (and whose title is in turn adapted from Archimedes), Bruno Latour steps back from *Laboratory Life*, written with Steve Woolgar four years earlier, and claims that

> it is time for sociology of science to show sociologists and social historians how societies are displaced and reformed with and through the very contents of science. But to do so, sociologists of scientific practice should avoid being shy and sticking only to the level of the laboratory.[58]

That is to say, artists and scholars embedded in scientific practice should expand beyond the 'microstudy', as Latour refers to his work with Woolgar on *Laboratory Life*. Decades later, Doing is disappointed that STS scholars have taken Latour's advice and neglected 'microstudies' that delve deeply into the production of specific facts. Doing would not be so disappointed if the practices of embedded bioartists were part of his analysis.

While artists do not necessarily consider the conventions of the STS laboratory study in their output, Catts and Zurr's reflections on their time in laboratories resemble microstudies that engage Doing and Latour's different approaches to laboratory space. In addition to focusing on discourse and discursive effects, the laboratory studies of artists emphasise the collaborative, interdependent nature of practice, and frame entire debates (as with *Disembodied Cuisine* and the question of *in vitro* meat). Artists expose the ways laboratories include and exclude the outside world by situating them in unexpected public places. Because artists bring practical skills learned in the laboratory to new situations, they enact laboratory studies, recombining the performances and performativities of laboratories with attention to the funding structures that make their work possible. The embed touches and transforms the substances surrounding her, revealing new connections and interdependencies by creating a critical contact zone.

7

Public Amateurism

Free Range Grain DNA Test Extraction Protocol

Prepare Sample

1. Take 15 grams of product (measure on electric scale with special paper)
2. Grind product

Extraction Protocol for Tepnel Biokits

1. Start with 10 grams of ground product
2. Add 30 ml of solution 1 (measure with cylinder) mixed with 1.2 ml of solution 2
3. Mix with Vortex
4. Incubate sample for 30 minutes in 65 C water (keep additional hot water around to maintain temperature).
5. Shake the samples every few minutes
6. Transfer 500 μl of solution into micro-tube (three samples are necessary)
7. Add 20 μl of Proteinase K
8. Incubate the samples for 30 minutes in water at 65 C. Shake regularly. (make a control sample for PCR – add all solutions but no product).
9. Cool on ice pack for 5 minutes
10. Add 54 μl of Solution 3 and invert tubes to mix (there can be no lumps)

11. Place samples back in ice pack for 10 minutes
12. Centrifuge at 8-9000 for 5 minutes
13. Merge samples (especially for processed food)[1]

~

This chapter considers material and rhetorical outcomes of the art laboratories of the Critical Art Ensemble (CAE), mapping histories of experimental life and activism alongside the sometimes unexpected consequences of public amateurism. CAE promotes amateur engagement with biotechnology, demonstrating that with just a few pieces of simple equipment, the curious consumer can determine the genetic makeup of their food. Such research may take place at home or in art galleries, supermarkets and laboratories, and can be adapted to different contexts. A quotidian, lived activism emerges from projects like *Free Range Grain* (2003–4) and *Molecular Invasion* (2002–4). This type of engagement echoes the rallying cry of Home Economics, that every task can be research. Everyday activities become research opportunities and possible avenues for policy shifts and education. Frameworks of action like 'Fuzzy Biological Sabotage' encourage communities to critically respond to agribusiness by reverse engineering test fields of GMO crops to make them vulnerable to the herbicide they have been designed to resist. CAE emphasises the do-it-yourself qualities of these interventions by empowering tactical positions and emphasising the impact that street-level interventions can have on larger industrial systems. By modelling amateur scientific investigation, CAE promotes a return to accessible experimentation not dominated by corporate interests. Such experimentation can be stimulating, as these practices involve participants in systems that would otherwise exclude them, but participation is not without risk.

In 2004, CAE co-founder Steve Kurtz awoke to find that his wife and CAE collaborator, Hope Kurtz, had died. Her death was the first chapter in a protracted legal ordeal involving Steve Kurtz, Robert Ferrell and their collaborators. Straying far from the initial tragedy of Hope Kurtz's death, the investigation of Steve Kurtz and Robert Ferrell resulted in the seizure of CAE's current projects, as well as sweeping changes in the way their work was displayed and the way future projects were executed. Perceptions of appropriate and inappropriate locations for biological materials and research animated these events. For Robert Mitchell, CAE,

Ferrell and Kurtz's legal ordeal marks the unfortunate beginning of a new era for bioart, an era 'in which new worries about the relative ease with which biological entities can be manipulated have led to changes in the social topology of exchange between biologists and artists'.[2] Bioterrorism becomes an increasing concern while biotechnical materials become more commonplace and transporting them becomes easier. The risks of learning in public increase. This chapter considers the consequences of public amateurism, tracking the perils and rewards of taking research and materials out of laboratory contexts and into domestic and public spaces.

After many years of producing 'tactical media' and public art interventions, in the late 1990s Critical Art Ensemble shifted their focus to biotechnology.[3] They see the confluence of biotechnology with capitalism as a problematic coupling in need of attention, education and creative resistance. Cultural neglect of biotechnologies could stem from the intimate ways in which we interact with these new technologies: we are often unaware that medicine and food, prescribed and purchased in private, are made from genetically modified organisms. To open up space for resistance within biotech worlds, CAE draws public attention to eating and reproductive bodies, specifically, bodies enmeshed in environmental degradation. In their essay 'The promissory rhetoric of biotechnology in the public sphere', CAE writes: 'now that humans have become a temporary set of biological relationships, an opportunity has arisen to redesign their biological matrix to better fit the needs of capital'.[4] Malleable human, animal and plant organisms may be easily ported to potentially exploitative contexts. Eugene Thacker considers the same terrain in his essay 'Data made flesh' when he observes that

> this tripartite process of encoding, recoding, and decoding the body operates through a kind of informatic protocol in which, at each step, information comes to account for the body. It is the process that I would like to refer to as 'biomedia'.[5]

Biomedia includes both the materials of biotechnology and the activities and composition of the eating body. Whether food, reproductive health or biowarfare, biomedia in the service of capitalism troubles CAE, first at the level of the reproductive human body,[6] and then across species boundaries and food systems.

Free Range Grain was produced in collaboration with Beatriz da Costa and Shyh-shiun Shyu. The installation assembles a mobile laboratory of several tables filled with equipment and laboratory paraphernalia. The project addresses changing GMO labelling legislation in the European Union, questioning the feasibility of labelling by asking after international policies on GMOs and doing practical tests to determine if food might have GMO ingredients. *Free Range Grain* presents an active workspace: power cords twine together under tables, supplies are stacked up around the larger machines and work surfaces are fully utilised, even crowded. Flanking the tables are computers where visitors can access the accompanying website and other resources. The mobile art laboratory tests foods supplied by gallery visitors for DNA markers that may reveal genetically modified ingredients. Suspect foods that test positive for possible genetic modifications are then labelled as such and offered for consumption.

Beatriz da Costa has speculated that the project would have been more resonant if located in a supermarket, but negotiating with retailers proved to be more difficult than negotiating with art spaces.[7] A grocery store setting would have emphasised how *Free Range Grain* engages with the ways in which test kitchens support the acceptance of new food technologies in the home. Test kitchens have strong ties to retail environments, from grocery store test kitchens developing recipes to promote their products to interactive kitchens providing samples and demonstrations in stores. *Free Range Grain* subverts the practical and ideological work of test kitchens to link laboratory protocols with disruptive political critique and public amateurism, building on home economists' call to make every project research. Inverting the ways in which test kitchens translate the potential of new food technologies to consumers, *Free Range Grain* generates critique rather than easy acceptance.

Free Range Grain was first exhibited in Germany and Austria amidst policy changes and debates about labelling GMO foods in the European Union. EU Directive 2001/18/EC repealed Directive 90/220/EEC to further clarify the EU's position 'on the deliberate release into the environment of genetically modified organisms'.[8] CAE's position paper on the

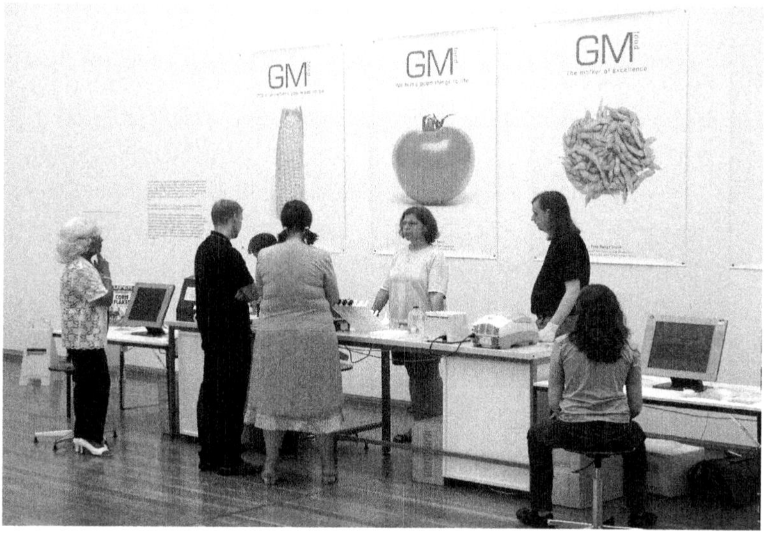

Figure 7.1 Critical Art Ensemble, Beatriz da Costa, and Shyh-shiun Shyu, *Free Range Grain* (2003–2004). Installation view, Schirn Kunsthalle, Frankfurt (2003). Courtesy of the Critical Art Ensemble.

project questions the European Union's ability to put this directive into practice, given United States policies on GM food:

> With the relatively heavy importation of grains and processed grains from the US, it appears to CAE/da Costa/Shyu that it will be very difficult to filter out GM foods. Not only are the protocols for product testing on a systematic basis different in intensity in every country, one has to also be suspicious about American ethical/legal resolve about volunteering information damaging to its profits. In the US, all the companies agree that labeling GM foods is not helpful to the public nor good for business.[9]

Here CAE points to practical differences between national policies involved in the global GM food economy, but targets the United States' 'ethical/legal resolve' rather than production processes. CAE locates impurities in United States policies, rather than in more tangible protocols. CAE subjects cultural theories to scientific protocols, asking how basic testing procedures might reveal leaky policies and permeable borders. *Free Range*

Grain accepts foods that are 'suspect', tests them, then offers them for consumption:

> People bring us foods that they find suspect for whatever reason, and we test them over a 72-hour period to see if their suspicions are justified. While we will not be able to say conclusively that a given food is genetically modified (although we can offer strong probability as [to] whether it is), we can test for conclusive negatives, and we can bring issues of food purity into the realm of public discourse. Identified contaminated food will be offered for consumption at your own risk [...] on the other hand, if we find nothing, it lends credibility to the idea that the European food chain is safe from GM contaminants.[10]

This process of identifying contaminated food for public consumption mirrors the EU's own regulations about labelling GM foods: 'The words "This product contains genetically modified organisms" shall appear either on a label or in an accompanying document.'[11] On a material level, both EU policies and *Free Range Grain* make a pile of 'identified contaminated food' to be 'offered for consumption at your own risk'. The performance illustrates how current policies are both fallible and, in the case of food labelling in the EU, impossible to enforce. New technologies wander, resisting containment and undermining our best efforts to sort, filter and control their flow.

With its frank, open design, *Free Range Grain* has an informal, even fun atmosphere. Previous CAE performances employed white coats, props and specially designed machinery to perform the science and business of biotechnology at a slight distance from participants. By contrast, *Free Range Grain* dresses casually: biotech systems are accessible, even messy. All of these machines are familiar to anyone who has taken university level biology classes or even classes in a well-appointed secondary school. Suspect food is mashed up, its DNA extracted from centrifuged pulp and then combined with enzyme digests before analysis by gel electrophoresis to reveal splits in the DNA, which might indicate genetic modification. With the goal of opening up laboratory work to non-specialists, the installation shows ordinary people doing ordinary things. *Free Range Grain* laboratories propose that everyone can be a researcher, and that the home and the supermarket are viable locations for molecular biology and critical inquiry.

Alongside *Free Range Grain*, a parallel project, *Molecular Invasion*, critiqued big food companies, Monsanto in particular, from an agricultural perspective. In *Molecular Invasion*, the Critical Art Ensemble, Beatriz da Costa and Claire Pentecost manipulate genetically modified cash crops designed to complement Monsanto's RoundUp herbicide. They lay the groundwork for 'Fuzzy Biological Sabotage' by developing an enzyme inhibitor to be deployed in fields where Monsanto crops grow. When RoundUp touches these plants, they die like their weed companions, thanks to 'reverse engineering' by CAE and 'rogue biologists'.[12] Both *Molecular Invasion* and *Free Range Grain* engage public perceptions of and concerns about GM food, and the differences between activist and consumer responses in the US and Europe. These plants are more than plants; they are political instruments, colonising forces, machetes carving up fragile ecologies.

In her essay 'What did you eat and when did you know it?,' CAE collaborator Claire Pentecost reconfigures the phrase 'from seed to shelf' as 'from dirt to dinner plate'.[13] In *Molecular Invasion* and *Free Range Grain*, CAE asks visitors to consider both the dirt and the dinner plate. The style of activist intervention CAE and their collaborators propose in *Molecular Invasion* resonates with actions in Europe located in the dirt, with less emphasis on the dinner plate. *Free Range Grain* focuses entirely on the dinner plate, examining consumption, packaging and labelling. In the late twentieth and early twenty-first centuries, consumers primarily focus on the dinner plate, especially in the United States, where the public expresses only mild concern about genetically modified foods and labelling is non-existent. European food activists focus on the dirt in the food consumption process, with protesters chopping down GM crops in the field. Nothing remotely comparable to Europe's anti-GM organising exists in the United States, where industrial monopolies are hardly threatened by manageable bourgeois counter-insurgencies like the slow food movement and the evangelism of Michael Pollan's best-selling food writing. Such rhetoric stands in for and quells the desire to act directly, providing a sanctioned framework for polite dissent while the 'eco-terror' vocabulary coined by conservative businessmen and politicians ensnares what anti-GM actions do occur, including crop and laboratory disruptions claimed by the Earth Liberation Front, Greenpeace and many local organisations.[14]

With *Molecular Invasion*, CAE acknowledges the thin line between the saboteur and the 'eco-terrorist' when suggesting tactics to deploy in fighting the 'molecular invasion'. *Molecular Invasion* and *Free Range Grain* create two very different kinds of art laboratories, with different openings and closures and different aesthetic and experimental goals. Like *Molecular Invasion*, *Free Range Grain* plays with the space between art and direct action, offering demonstrable results about the food people buy at the supermarket, but the performances manipulate food and eating from different locations in the food supply web. *Free Range Grain* stays close to the supply chain, while *Molecular Invasion* considers the agency of plants that become raw materials in industrial food systems. In addition to the human performers addressing agribusiness corporations, the Roundup Ready crops' performances of vulnerability contribute to *Molecular Invasion*'s legibility. Do audience members eat the crops in the gallery? Probably, if indirectly. Most cash crops are consumed after being processed into feed for livestock, ethanol, starch-based plastics or high fructose corn syrup. Plants become symbolic markers, signifying sites of new food technologies. Do audience members eat the products tested in *Free Range Grain*? Almost certainly. Everything in the gallery was sourced from a local grocery store or market. With its immediate and accessible commodity chain, *Free Range Grain* feels less like fighting or sabotage and more like a new domestic activity to help us visualise the flow of GMOs, in the tradition of the quotidian research practices of Home Economics. What if the PCR (polymerase chain reaction) machine became a kitchen appliance? *Free Range Grain* suggests that testing for DNA splices might one day become a routine kitchen task, gathering data points from the linoleum up.

CAE wants this type of quotidian critical inquiry to become the norm, making grocery shoppers into public amateurs. How does the role of 'public amateur' function in bioart practice? Just as laboratories function in many different ways in bioart projects, so too do bioartists occupy a range of positions within the laboratories they visit and those they create themselves. Each laboratory allows for different perspectives and relations between laboratory work and art work. Two interrelated models of artist/scientist collaboration encompass many possible configurations and relationships. One model is the 'embedded' artist collaborator, who usually situates themselves in an established laboratory.[15] CAE prefers a different model, that of the unsanctioned public amateur, 'a person who consents to learn in public'.[16] Claire Pentecost theorises public amateurism as:

> active social participation in which any nonspecialist is empowered to take the initiative to question something within a given discipline, acquire knowledge in a noninstitutionally sanctioned way, and assume the authority to interpret that knowledge [...] the motive is not to replace the specialist, but to augment specialization with other models that have legitimate claims to producing and interpreting knowledge.[17]

The learning curves of bioart and workshops like those conducted by SymbioticA produce public amateurs, inviting Pentecost's 'other models' into the field of practice already occupied by biological science, corporate interests and academic disciplines. Amateurs are not connected to institutions; instead, they create their own spaces of production and experimentation. Often the amateur works in the home, conducting experiments that might be made public through publishing. Public amateurs find ways to validate their work within larger systems of knowledge, which may call on or contribute to the expertise of institutionally sanctioned experts. They do not view amateurism as a step towards eventual professionalism, but prefer to promote 'other models' as autonomous, if different, ways of producing knowledge.

Public amateurism ultimately contributed to Steve Kurtz's legal ordeal, which began with confusion about the appropriate place for and use of biological materials. Gregory Sholette shares this view, writing that 'CAE's avowed amateurism [...] may also be playing a central role in the group's recent predicament with Federal authorities'.[18] *Free Range Grain* enacts the globalisation of food markets and the impossibility of perceiving, let alone labelling, genetically modified foods, yet this is one of CAE's milder, friendlier installations. With its sharp rhetoric of Fuzzy Biological Sabotage, *Molecular Invasion* would appear to be the more challenging and potentially controversial installation. But *Free Range Grain* happened to be awaiting shipment when Hope Kurtz died and agents from multiple federal agencies occupied the Kurtz home. Before Hope Kurtz died, *Free Range Grain* was a public demonstration of laboratory processes related to genetically engineered food. This project had none of the militant undertones of

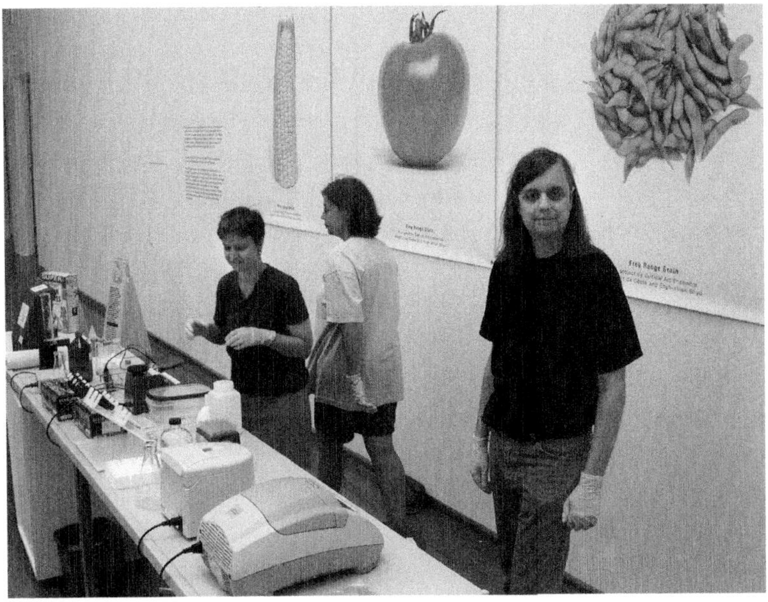

Figure 7.2 Steve Kurtz in the *Free Range Grain* laboratory. Schirn Kunsthalle, Frankfurt (2003). Courtesy of the Critical Art Ensemble.

Molecular Invasion and was conceived of in educational terms as a way of promoting amateur scientific investigation. After Hope Kurtz died, a number of limitations were placed on CAE's practice. Bacteria used in CAE's newest work, *The Marching Plague*, were conflated with the *Free Range Grain* materials seized by the FBI. CAE was subsequently asked to avoid displaying any biological agents in US exhibition spaces (notably, during the 2006 Whitney Biennial, the museum insisted that CAE limit their *Marching Plague* installation to videos of actions performed in Europe[19]).[20] A grand jury was convened. Charges that began with bioterrorism were reduced to mail and wire fraud – charges that carry sentences of up to 20 years, thanks to the PATRIOT Act.

Because *Free Range Grain* materials were present when Kurtz's belongings were seized, its content and methodology were often perceived as being connected to this early bioterrorism charge. This misreading connects the dots between bioterrorism and the prolific genetic engineering practices that alter global food supplies – an unintended but prescient outcome of

Kurtz's legal battle. *Free Range Grain*'s critique of GM foods and related legislation falls away from post-2004 conversations about the project. This could be because many written accounts of Kurtz's ordeal paid little to no attention to CAE's work and installations, focusing instead on CAE's storage of their project materials in Kurtz's home and subsequent interactions with authorities. It could also be because promoting CAE's written and performance projects as critical inquiry would somehow lend credence to the FBI's investigation and the federal lawsuit against Kurtz and Ferrell. What readings were possible before Hope Kurtz died, and what readings are we left with now, after CAE displayed an empty table while a contingent of supporters waved homemade protest banners at the Massachusetts Museum of Contemporary Art?[21] How can CAE desire to 'escape the terrorist label altogether' while simultaneously choosing to practise and speak as renegade, tactical saboteurs engaging in a fight against invasion? Nicola Triscott raises the concern that CAE's decision to turn away from biotechnology in recent work 'may imply that the US Department of Justice were, to some extent, successful' in their efforts to silence CAE's critiques of big food.[22]

Writing about 'the post-ecologist condition', Bronislaw Szerszynski emphasises CAE's use of 'complex forms of communicative irony'.[23] Their performances prompt 'observers to work harder to recognise that there is a hidden meaning beneath, and in stark tension with, the overt meaning', producing 'irony-as-world-relation'.[24] This tactical irony, at least in recent projects, emerges in part from taking plants seriously as actors; the work relies on double meanings and multiple interpretations to reveal the limits of the category human when working with GM food and biotech research. The individual human subject has no place in the political landscapes CAE maps and critiques. Instead, collective forms – alliances between humans, non-humans, machines and organisations – emerge as the agential forces that produce world relations. The US Justice Department's case against Steve Kurtz reveals a disconnect between the practical activisms on offer when visiting a CAE art laboratory and the capacity of governmental agencies to interpret communicative irony.

The case against Kurtz and Ferrell was built around materials perceived as appropriate or inappropriate in a range of locations, the most important of which is the Kurtz home. The case also highlights the importance of performance and mimesis. Writing about several artists in the US who have been arrested or detained since 11 September 2001, Oliver Watts observes that 'the Government's authority, both as panoptic and military machine respectively, was parodied by these artists so closely that they became confused with the enemy other, the terrorist'.[25] By attempting to both address and escape the label 'terrorist', Kurtz and others trespass on the State's 'position as the only apparatus that may name the threat of terror'.[26] The case against Kurtz shows how CAE's publishing and creative acts conflict with a judicial reality that refuses concepts CAE's work presupposes: productive dissent as a constitutive force in democracy, the need for practical art utopias, for oppositional spheres, and the possibility that scientific objects can move freely between art spaces, laboratories and studios.

Oppositional spheres manifest outside of installations as well: letters of support and science publication editorials devoted to the Kurtz/Ferrell case highlight their collaboration and the way that universities produce research.[27] Speaking through actor Peter Coyote in Lynn Hershman-Leeson's film *Strange Culture* (2008), Robert Ferrell reflects on his ordeal:

> I don't think I did anything wrong. I would do the same thing again with Steve or anyone, if they wanted reagents and they were serious people that convinced me that they were doing all this for educational reasons. That's the whole point, to educate the public about what is going on in science today. If I've broken any laws as far as Steve is concerned, I've broken it a thousand times, because I've had many people in my lab to learn these techniques and given biological reagents to many students to work with.[28]

Given his deteriorating health (he suffered both cancer and a series of strokes in the four years between the initial FBI investigation and Kurtz's acquittal), Ferrell eventually pleaded guilty to a misdemeanour charge, 'mailing an injurious article', for which he received a $500 fine and no jail time.[29] Support for Ferrell and Kurtz has come from artists and researchers who share knowledge and samples on a daily basis and who believe the university to be a space capable of generating oppositional thinking. Before this case, academics, artists and researchers had been warned in other ways, but many trace the beginning of post-PATRIOT Act self-censorship across the arts and humanities to this case in particular.[30] Anna Munster connects the case to the FBI surveillance practiced under McCarthyism, the NEA 4 and the '"chance" harvesting of illegal mainly Middle Eastern, Indian and Pakistani immigrants'.[31] The Kurtz case also resonates with the long history of targeting activist organisations, given CAE's collective knowledge distribution practices. Among activists, group identity eclipses individual speech.[32] This is also true of the corporations that engineer GM food, where corporate entities play abstract 'persons' for legal purposes. CAE's projects extend the symbiosis of space and association within the museum exhibition, producing laboratories that force us to acknowledge our complicity with corporate structures and those structures' complicity with deep histories of colonial economies and capitalist co-optation of bio-media. Beyond experimental encounters with cultural data, CAE's laboratory spaces present the work of collectivity.

CAE's books and installations encourage their audience to become participants. Their work joins projects like Natalie Jeremijenko and Heath Bunting's *Biotech Hobbyist* magazine, which proposes a curriculum of activities and resources designed to empower individual producers by 'recognis[ing] that some of the greatest cultural and technological advances have emerged from people's bedrooms'.[33] Like *Biotech Hobbyist*, CAE's books and installations outline simple ways to practice anti-corporate research at home. By modelling and invigorating amateur scientific investigation, CAE precipitates a return to accessible experimentation not dominated by corporate interests. Such experimentation generates discourse and produces philosophy and political action, involving participants in systems that would otherwise exclude them. Rebecca Schneider and Jon McKenzie ask 'whether the "hands-on" quality of CAE's work has triggered the charge of bioterrorism, not their use of biological substances. Is CAE's general move away from *rhetoric* toward tactile *participation* pointedly what provoked the FBI to suspect terrorism?'[34] Although Schneider and McKenzie's reading gives the FBI perhaps undeserved credit for thoughtfully reading performance art, they identify a threatening element in CAE's projects, which work by empowering tactical positions and emphasising the impact that street-level interventions can have on larger industrial systems.

Provoking antagonism models a way of resisting corporate control that acknowledges mutual responsibility: CAE challenges Monsanto's agenda by engaging their process rather than burning their product.

Anna Munster connects the way the case was pursued with networked surveillance technologies, implying that Kurtz's case could ripple across networks to change how future cases are targeted and assessed. This theory predicts Mitchell's third era of bioart, 'Bioart and bioterrorism', and some of its possible consequences.[35] Munster writes:

> once CAE had been entered into the FBI's managed information territories and mapped them as a hub, their practices automatically figure as nodes to be linked in the informatic conduct of the war on terror [...] The fraud charges become a way of weakening the links that CAE hold with the scientific community, as they implicate a collaborative relation between Kurtz and a geneticist.[36]

Collaboration itself ends up being policed by way of collaborative systems, with complex networks and human–machine alliances providing the mechanisms by which overlaps between art and science might be monitored and undermined.

Hershman-Leeson's *Strange Culture* was one of many cultural productions to emerge in response to the case against Kurtz. The film addresses wandering materials, collaborative research and Mitchell's 'third era of bioart' through interviews, re-enactments and excerpts from CAE's documentation. Hope Kurtz appears in *Strange Culture* in several different ways. Tilda Swinton enacts her last days at home with her husband Steve (played by Thomas Jay Ryan). She appears briefly in photographs as a young woman, distant from the Hope portrayed by Swinton. We learn things about her from Steve Kurtz throughout the film: she 'was always a huge fan of Tilda Swinton', she had a 'very uncanny genius for pattern recognition'. And, early in the film, we learn that Hope Kurtz died of heart failure on 11 May 2004. In their essay 'A feminist perspective on bioterror: from anthrax to Critical Art Ensemble', Stephanie Kane and Pauline Greenhill examine Hope's body in the context of sexual and bio-terrorism. State response to Steve Kurtz undercuts the lack of resources devoted to sexual violence:

> The body of Hope Kurtz has lessons of its own. Used as a pretext for what was once unlawful government action, then quickly superseded by state concerns with war, Hope Kurtz's corpse is a map of the contours of protection afforded women by a masculinist militarized state.[37]

While domestic violence was not a factor in her death or in the investigation of Steve Kurtz, Kane and Greenhill point out that sexual violence and the death of female bodies has been normalised and ignored in favour of questions of war, reducing Hope Kurtz's death to a 'pretext'. Most critics and journalists writing about the case have ignored her body's role in her husband's subsequent prosecution. A newscast delivered by Keith Olberman opens the film, bearing out Kane and Greenhill's hypothesis that authorities overlook violence against women:

> Like many an unfortunate drama, the story begins with a death. Steve Kurtz called 911 early on the morning of May 11, after his wife suffered cardiac arrest and died in her sleep. When police arrived on the scene, they saw not a 45-year-old woman claimed before her time, but rather petri dishes and sophisticated scientific equipment.[38]

Hope's body is invisible, unseen; instead, police see materials that have strayed into inappropriate places. In *Strange Culture*, perhaps thanks to Steve Kurtz's involvement, we at last see Hope Kurtz as more than an unfortunate starting point for the bioterrorism case against her husband. Throughout the film, Kurtz and others, including Steve Barnes, Steve Dietz, Beatriz da Costa, Claire Pentecost and Nato Thompson, describe how what they saw differed from what the police and FBI saw. Their inability to see CAE's laboratory equipment and materials as the government did – as dangerous and suspect – led CAE and their collaborators to believe that eventually government agencies would realise they had made a mistake.

After paramedics and police entered the Kurtz's home, a long chain of bureaucratic reactions was set in motion. *Strange Culture* narrates these events using a combination of actors, interviews, actors portraying themselves, re-enactments and illustrations created by Timothy Stock and Warren Heise from their graphic novel *Suspect Culture: An Exposition of the Death of Hope Kurtz*.[39] The illustrations show events that Kurtz could not recount due to legal restrictions and, coupled with an eerie score by the Residents, lend the film a tense, claustrophobic atmosphere. Hershman-Leeson builds layers of what Kurtz calls 'hyper-reality', collapsing then separating elements of the case. With its shifting viewpoints and multiple personalities, the film emphasises the mutability of media and perception. While a curator sees the Arabic script in an Atlas Group work as an uncontroversial addition to an exhibition publicity flyer, a FBI agent sees the same script as evidence of a possible association with a terrorist organisation. Portrayed as interpretable and unstable, laboratory materials figure prominently both in Kurtz's actual home in Buffalo and in a recreated set in San Francisco, as do CAE's politics and methodology. CAE's amateurism, publishing practices and engagement with words like 'saboteur' and 'terrorist' combine to reveal a shrinking space for critical work about sensitive topics. Other contemporary cases in the United States, including those of plague researcher Thomas Butler, art student Clinton Boisvert and Massachusetts Institute of Technology student Star Simpson, involve wandering materials, transgressive placement of charged objects and procedural misfires similar to those cited by the FBI in the Kurtz case.[40] Fortunately for Kurtz, charges against him were dismissed in 2008, four long years after Hope Kurtz died.[41] Robert Ferrell, Butler, Boisvert and Simpson were all charged with crimes as a result of materials escaping, being misplaced or being seen in the wrong place at the wrong time.

Clearly the PCR machine cannot yet be safely deployed as a kitchen appliance – at least not in the United States. *Free Range Grain* proposes that the DNA extraction protocol that prefaces this chapter will be accessible to average consumers and may become an everyday practice in kitchens of the future. *Strange Culture* stays in the present, asking how materials can be perceived in radically different ways from one person or place to the next. The film illustrates how the case against Kurtz shuts down speculative possibilities, leaving us with paranoia and concern regarding what materials we put where. While projects like *Free Range Grain* and *Molecular Invasion* open up potential futures, proposing alliances and actions around topics that are currently undertheorised, judicial response to the material semiotic praxis fundamental to these projects forecloses those same speculative potentials. None of *Strange Culture*'s nested hyper-realities express the breadth and idealistic expansiveness of

CAE's work before Hope Kurtz's death in 2004. In their pursuit of participatory dissent, CAE's changing practice reveals a tightening space for imaginary futures and present-day antagonism within American democracy. Rhetorical choices and material traces move to the fore, while celebrations of the ephemeral and the speculative disappear. If we limit our ability to make such connections to prescribed spaces and if the label 'art' ceases to protect those spaces, then democratic gestures are both lonely and dangerous.

8

Cookbook

Garrulous Gazpacho

5 or 6 large ripe mouthy tomatoes
1 cucumber
2 cloves garlic, minced
1 tbsp lip lemon juice
1 tbsp red wine vinegar
2 tbsp chopped fresh basil
¼ cup extra-virgin olive oil
1 package (8 oz) earsprouts
salt and pepper
½ tsp Tabasco sauce (optional)

1. To prepare the mouthy tomatoes, soak in mouthwash for ten minutes, then rinse in clear water.
2. Reserve the mouths for garnish, and finely dice or roughly food process the flesh of the tomatoes and the cucumber.
3. In a large bowl, combine the diced tomatoes and cucumbers, the minced garlic, the lip lemon juice, the red wine vinegar. Stir in the olive oil and the fresh basil. Add salt and pepper to taste.

4. Chill thoroughly, about 3 hours or overnight.
5. Serve cold and garnish with the reserved mouths. Immediately before serving, sprinkle liberally with earsprouts.[1]

~

Cookbooks circulate commercially as well as in activist and art contexts to both promote and critique new food technologies. While promotional materials tend to be unobtrusive (a recipe on the side of a food package, recipe cards packaged with coupons in the mail), recipes and cookbooks from activists and artists are often critical, provocative and humorous. Cookbooks are a favourite vehicle for critiquing new technologies like genetically modified foods. Recipes make often-invisible genetically modified ingredients visible and give our interactions with them a sense of urgency. Recipes have narrative force; we visualise ourselves following the directions as we read. Our imaginary dicing, chopping and blending make techno-foods real.

A range of earnest 'No GMO' cookbooks and lifestyle blogs work to document known genetically modified organisms and offer alternatives.[2] Most of these critiques overlook biofortified crops like golden rice or virus resistant GMOs like ringspot resistant papayas, instead appealing to wealthy consumers who seek to return to a mythical 'before' when our food supply was supposedly untainted by genetic modifications. Anti-GMO cookbooks promote purity, seeking to protect and conserve food ecologies from perceived bioengineered threats. At the other end of the spectrum, immunologist Beda Stadler presents an unusual pro-GMO cookbook titled *Genes on the Fork*. The recipes include genetically modified ingredients but emphasise that not all genetic modifications are the result of DNA splicing.[3] GMO advocates often connect current practices back to Mendel's peas, arguing that new technologies simply extend old, safe techniques. As comforting as such a philosophy may be, selective breeding in the greenhouse only succeeds within a range of compatible species. These traditional methods will never produce the inter-kingdom hybrids that emerge from biotechnology laboratories and are sometimes labelled 'Frankenfoods'.

Greenpeace produced *Good Food Gone Bad* in 2010. The cookbook follows apron-clad José Manuel Barroso, President of the European Commission, and John Dalli, former European Commissioner for Health and Consumer Policy, as they prepare dishes like Angry Farmer Antipasti and Corporate Control Tart.[4] The conceptual ingredients in the recipes include Corporate Domination pastry, freshly chopped Marketing Hype, and a toasted tortilla of Rural Poverty. Each recipe's ingredients form a short fact sheet, complete with citations. The cookbook functions as a pointed critique of specific European Union policies and practices: at the time of its creation, Barroso and Dalli had recently approved the licensing of genetically modified Amflora potatoes. Amflora potatoes yield a waxy starch that can be used in papermaking and adhesive manufacturing. Even though Amflora potatoes are not for direct human consumption (the starch has been approved as animal feed) and are grown only for industrial applications, the cookbook format allows for a wide-ranging critique of GM crops. Amflora's life in the EU was short – production was stopped in 2012. Also in 2012, Barroso asked Dalli for his resignation amidst fraud allegations, so the two no longer share a kitchen.

These cookbooks from GMO activists and advocates provide a backdrop for artistic engagements with the cookbook form within the GMO debates. This chapter reads three such cookbooks, Christine Chin's *The Genetically Modified Foods Cook Book*, the Critical Art Ensemble's *Betty Crocker 3000 Presents Food for a Hungry World* and Next Nature Network's *The In Vitro Meat Cookbook*. As with *Good Food Gone Bad*, these projects take up the cookbook format as a vehicle for critique. Second-wave feminists hesitated to study or celebrate cookbooks, finding them oppressive and confining. Decades later, the artists and activists in this chapter acknowledge the cookbook format as a generative path into controversial topics like genetically modified foods. Contemporary feminist scholars have also embraced a critical interest in cookbooks, with a range of scholarly articles and books on cookbooks and recipes published in recent years.[5] The increasing number of cookbooks written as critical texts indicates that while women remain our favourite recipe keepers and cookbook authors, feminine forms are no longer perceived as an obstacle to making serious contributions across a range of public debates.

Questions about visibility animate artistic engagements with GMOs and new food technologies. How might transspecies genetic modifications be made visible in our food supply? And how do visual cues hint at the wider sensory worlds unfolding within these new technologies? Labelling in supermarkets continues to be the subject of legislation and debate but labelling does not interrogate or reveal the technologies that animate genetically modified foods. Skipping the labelling debate in favour of speculative material encounter, artist Christine Chin created a series of sculptures called *Vegetable Human Hybrids* (2003–5). The hybrids include oranges with orifices, limes with teeth, squash with beards and 'toetatoes', potatoes shaped like large, knobby toes.[6] Chin developed *The Genetically Modified Foods Cook Book* (2005) and related cooking show as 'a way to insert these hybrids into our daily lives'.[7] Chin sees the cookbook as a way to connect with quotidian elements of our lives and kitchens while simultaneously questioning transgenic food technologies.

The cover of *The Genetically Modified Foods Cook Book* reinforces this everyday aesthetic, quoting popular binder-style cookbooks like Betty Crocker's *Picture Cook Book* and the chequered tablecloth pattern covering *Better Homes and Gardens New Cook Book*. Betty Crocker's cutting edge *Picture Cook Book*, also known as Big Red, set the bar for future binder cookbooks' graphic design. Recalling Betty Crocker's layout, Chin includes step-by-step captioned photographs to accompany larger images of the finished dish. Although Chin stops short of embodying the corporate mascot format pioneered by General Mills with Betty Crocker, *The Genetically Modified Foods Cook Book* does play on the cookbook aesthetics and commercial agendas developed by General Mills and *Better Homes*. In addition to the cookbook, 'Chef Chin' hosts a cooking show available on DVD.[8] As in Home Economics demonstrations of the past, the cooking show introduces techniques for working with new foods and new technologies. Chin treats each dish as a sensory and social system. Her garrulous gazpacho pairs with earsprouts, which listen to the mouthy tomatoes. Chin asks how a cookbook can illustrate the cross-species shifts occurring in our food supply and how multiple species are at play in any given meal interacting with one another. Recipe artists comment on the systems that sustain new food technologies and the ruptures that occur when old habits collide with new foods. What does a tomato expressing human proteins for vaccine development look like? In Chin's cookbook, these imaginary products blur species boundaries as well as boundaries separating genres, the real and the fictional.

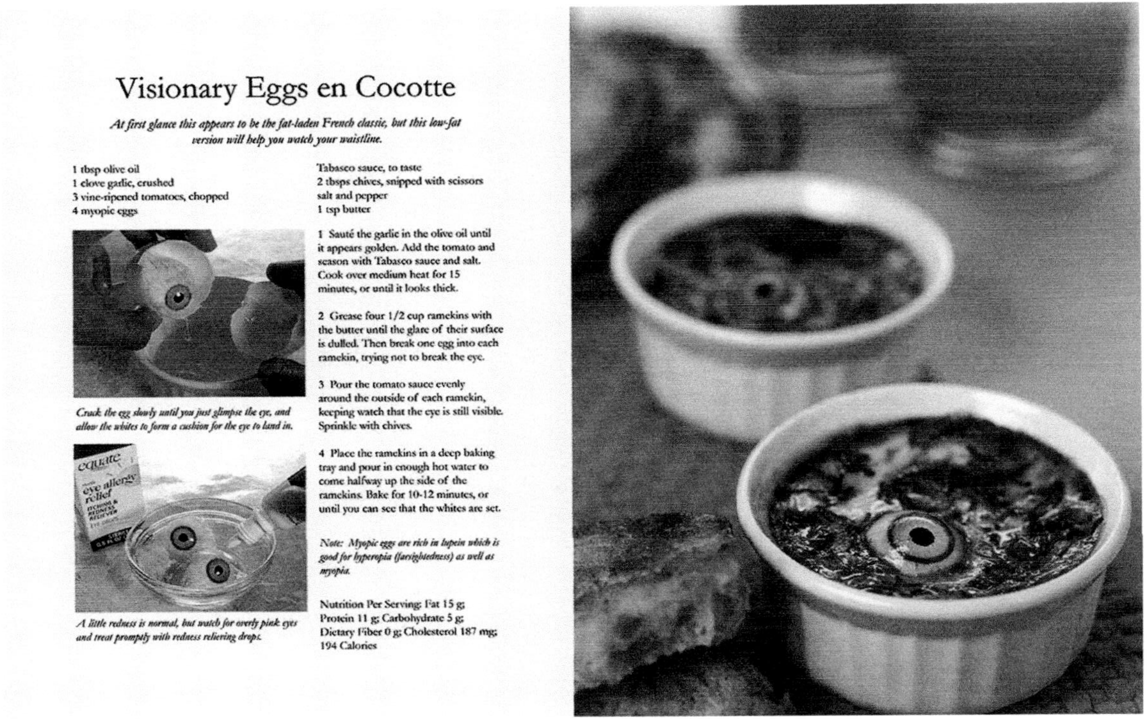

Visionary Eggs en Cocotte

At first glance this appears to be the fat-laden French classic, but this low-fat version will help you watch your waistline.

1 tbsp olive oil
1 clove garlic, crushed
3 vine-ripened tomatoes, chopped
4 myopic eggs

Tabasco sauce, to taste
2 tbsps chives, snipped with scissors
salt and pepper
1 tsp butter

Crack the egg slowly until you just glimpse the eye, and allow the whites to form a cushion for the eye to land in.

A little redness is normal, but watch for overly pink eyes and treat promptly with redness relieving drops.

1 Sauté the garlic in the olive oil until it appears golden. Add the tomato and season with Tabasco sauce and salt. Cook over medium heat for 15 minutes, or until it looks thick.

2 Grease four 1/2 cup ramekins with the butter until the glare of their surface is dulled. Then break one egg into each ramekin, trying not to break the eye.

3 Pour the tomato sauce evenly around the outside of each ramekin, keeping watch that the eye is still visible. Sprinkle with chives.

4 Place the ramekins in a deep baking tray and pour in enough hot water to come halfway up the side of the ramekins. Bake for 10-12 minutes, or until you can see that the whites are set.

Note: Myopic eggs are rich in lupein which is good for hyperopia (farsightedness) as well as myopia.

Nutrition Per Serving: Fat 15 g; Protein 11 g; Carbohydrate 5 g; Dietary Fiber 0 g; Cholesterol 187 mg; 194 Calories

Figure 8.1 Christine Chin, 'Visionary Eggs en Cocotte' from *The Genetically Modified Foods Cook Book* (2005). Courtesy of Christine Chin.

The Genetically Modified Foods Cook Book presents strategies for integrating what Chin calls 'human-vegetable hybrids' into daily life. Of her work, Chin writes:

> The basic idea behind the *Cook Book* is to make an invisible technology visible. GM corn looks just like conventional corn, despite the addition of bacterial genes that aren't even from the plant kingdom. But in the invented world of the *Cook Book*, a carrot with human genes looks like a carrot with human genes[9]

Chin alludes to her 'Finger Rolls' recipe, which calls for 'finger carrots': regular carrots that have grown human nails and cuticles. In her recipe for 'Finger Rolls', the finger carrot peeks out from behind other ingredients, tastefully arranged on a bamboo mat. The material manifestation of something human in your food returns throughout the cookbook:

> Cannibalism […] is a favorite topic. To some extent, I use it shamelessly as part of what keeps people interested […] the yuck factor of finding a human part in your food, or of chopping up a human part. But it also alludes to some of the messy implications of what happens when you start recombining genes[10]

By consuming genetically modified organisms, we are potentially consuming and cannibalising ourselves. While feminist artists consumed themselves in a metaphorical, political sense, returning abstract, energetic sustenance to the giving and often depleted feminine sphere, Chin's human-vegetable hybrids suggest that we are consuming ourselves in a fleshier, more literal way. Chin's cannibal commentary also touches on more general anxieties provoked by 'human parts' in our food and cross-kingdom genetic modifications. Should animal and bacteria cells travel into plants? How do such organisms undo taxonomic systems? What do we eat when we eat recombinant technologies and how does our incorporation of these organisms alter our own composition?

Human tissue has, of course, appeared in our food before the advent of genetic engineering. Most infamously, Münsterberg shopkeeper Karl Denke sold human flesh as 'smoked pork' in 1920s Germany.[11] 'Survival

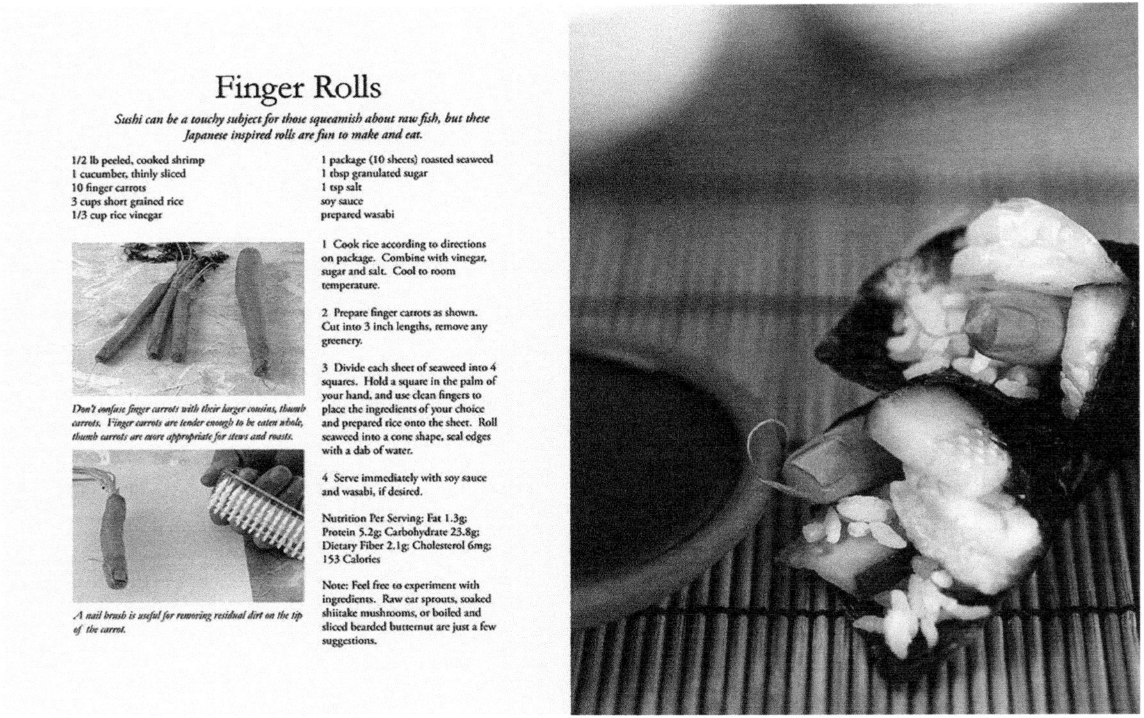

Finger Rolls

Sushi can be a touchy subject for those squeamish about raw fish, but these Japanese inspired rolls are fun to make and eat.

1/2 lb peeled, cooked shrimp
1 cucumber, thinly sliced
10 finger carrots
3 cups short grained rice
1/3 cup rice vinegar

1 package (10 sheets) roasted seaweed
1 tbsp granulated sugar
1 tsp salt
soy sauce
prepared wasabi

Don't confuse finger carrots with their larger cousins, thumb carrots. Finger carrots are tender enough to be eaten whole, thumb carrots are more appropriate for stews and roasts.

A nail brush is useful for removing residual dirt on the tip of the carrot.

1 Cook rice according to directions on package. Combine with vinegar, sugar and salt. Cool to room temperature.

2 Prepare finger carrots as shown. Cut into 3 inch lengths, remove any greenery.

3 Divide each sheet of seaweed into 4 squares. Hold a square in the palm of your hand, and use clean fingers to place the ingredients of your choice and prepared rice onto the sheet. Roll seaweed into a cone shape, seal edges with a dab of water.

4 Serve immediately with soy sauce and wasabi, if desired.

Nutrition Per Serving: Fat 1.3g; Protein 5.2g; Carbohydrate 23.8g; Dietary Fiber 2.1g; Cholesterol 6mg; 153 Calories

Note: Feel free to experiment with ingredients. Raw ear sprouts, soaked shiitake mushrooms, or boiled and sliced bearded butternut are just a few suggestions.

Figure 8.2 Christine Chin, 'Finger Rolls' from *The Genetically Modified Foods Cook Book* (2005). Courtesy of Christine Chin.

cannibalism' occurs on ill-fated outdoor explorations, during famines and after plane crashes. Vaccination can be thought of as cannibalistic: Priscilla Walton argues that vaccines fit the definitions of cannibalism when we think of shots as 'treated blood or tissue matter from one source, often human, injected into another and consumed in various ways for the host's physical maintenance'.[12] Cannibalism regularly appears in science fiction (1973's *Soylent Green* features processed and rationed 'Soylent Green' wafers made of 'people' and distributed to an overpopulated future world where strawberry jam costs $150 a jar), and in human illnesses like anorexia, which Walton suggests is akin to 'cannibalizing the self'.[13] Cannibalism is not an exclusively human problem. Michael Pollan and Walton both investigate how cannibalised livestock feed has resulted in disease.[14] American cattle continue to eat each other, even after meat and bone meal were banned from US livestock feed in 1997: the ban makes an exception for blood and fat.[15] Elements of all of these cannibalisms coexist in Chin's work, with a side helping of comic relief. Like Chin, writer Ray Tannahill connects cannibalism and humour, observing that 'humanity's obsession with the symbolism of flesh and blood has taken some irresistibly comic turns over the centuries'.[16] Whether vampiric, pornographic or futuristic, the cannibal is 'within all of us, should the circumstances demand the appearance'.[17] Choosing cannibalism as a way to frame transspecies genetic engineering addresses the interdependence of humans and food supplies, while simultaneously asking after the muddled lines between animal, plant and bacteria.

Chin's work focuses on human senses – vegetables and fruits have eyes, ears, fingers and teeth. Anti-GM activists also take up this strategy when they anthropomorphise vegetables and animals, especially laboratory rats and cows. In addition to the humorous creepy feeling Chin describes, body parts compel visceral identifications with hybrid forms. Transspecies alliances make both real and fictional GM vegetables possible, and Chin marks those alliances at the level of human and vegetable skin. In contrast to the anti-GMO cookbooks advocating a return to a lost moment when food and the natural world harmoniously coexisted, recipe art provides a critical alternative, asking how we might reinterpret our kitchens, cookbooks and ingredients as political vehicles. We have come to understand how the personal human life can be political, but the private lives of plants, livestock and Twinkies are only now entering the political realm. Chin brings industrial food worlds to the dinner plate, writing scripts that must be enacted without relying on a safe distance between the private kitchen and the industrial systems responsible for current and future outrageous ingredients. This critique not only addresses efficiency, gendered labour and domesticity, but also forces the complexity of industrial and GM food production into the smaller world of the private kitchen, prompting a confrontation between polluted GMO food networks and ordinary kitchens which makes recourse to 'the integrity of natural kinds' impossible.[18]

The In Vitro Meat Cookbook embeds itself in many of the same ethical entanglements that preoccupy Chin. From kitchens of the future to celebrity and self-cannibalism, the cookbook gathers current and speculative designs circulating in the world of *in vitro* meat. Some of these processes and products will be familiar from Chapter 6, 'Meat Culture' in this volume (the Postburger and the Tissue Culture and Art Project's *Disembodied Cuisine*), while others are entirely new, invented for this 'cookbook from which you cannot cook'.[19] Cannibalism becomes an edible outcome of celebrity 'culture' with a recipe for 'Whiskey-Glazed Celebrity Cubes', while 'In Vitro Us' roasts tiny 'effigies' nourished in a 'personal bioreactor worn as a pendant'.[20] The cookbook also focuses on commercial and domestic kitchen appliances, with experimental bioreactor implementations like the 'In Vitro Kebab' and the 'Home Incubator'.

The book's authors are affiliated with Next Nature Network, located in Amsterdam, and with the Industrial Design department at Eindhoven University of Technology. Their sponsoring organisations include New Harvest, also based in the Netherlands. This diverse group of collaborators interrogates every step of meat production and consumption with the aim of thinning the 'thick interface separating the experience of eating from the process of food production'.[21] Recipes and technologies are given a star rating system, with five stars indicating that a given dish could go into production using technologies available today. ('Celebrity Cubes' have four stars; 'In Vitro Oysters' have only one.) Like Chin's *Genetically Modified Foods Cook Book*, these recipes bring technology to the table with humour and careful attention to fragile distinctions between human and animal meat.

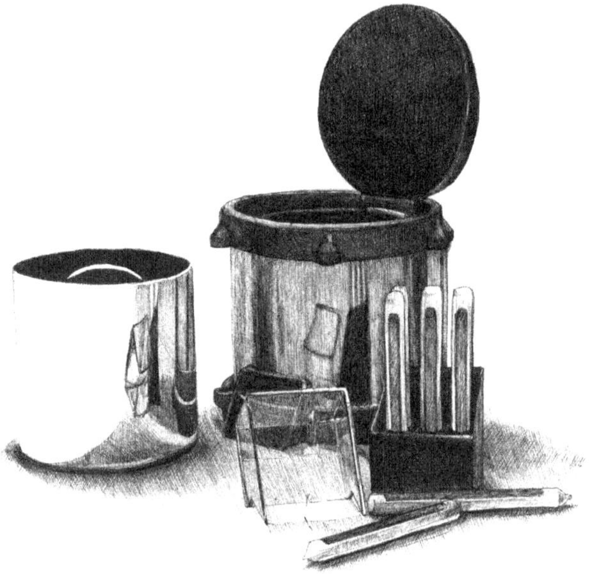

Figure 8.3 Next Nature Network, 'Home Incubator' from *The In Vitro Meat Cookbook* (2014). Concept design: Daniel Ong. Illustration: Silvia Celiberti.

In vitro meat democratises cellular origins – these meats could be cultured from any organism, even synthetic ones. To counter the possibility that flattening distinctions between species might threaten the positive interactions humans and animals maintain, even in a farming context, Next Nature Network imagines a new form of animal husbandry, the 'Pig in the Garden'. Proposing that a cooperatively kept pig might become 'a living reservoir of stem cells to grow in vitro meat', the cookbook imagines a social scenario where 'the pig itself could become a beloved ambassador of the community'.[22] The pig serves as a symbol and a transitional object to help communities learn to regard animals differently. As with the distributed monarch in Julian Huxley's 'Tissue culture king', the pig's circulating cells multiply a bond between people. The pig in the garden defines the garden and the group of humans feeding off its flesh. Other animals might stand in for the pig, forming a material semiotic bond among groups who do not eat pork: one might imagine a chicken in the garden or a fish in the pond. Keeping the animal in the garden's cells stable in a domestic context involves appliances like the 'Home Incubator' and the 'In Vitro Aquarium'. The speculative designs in *The In Vitro Meat Cookbook* manage to side-step concerns about 'big food' and the possible ways that a few huge conglomerates might still monopolise future eating. Instead, these recipes and devices stay local, dwelling in the home rather than the factory farm. A system of pigs in the garden and home bioreactors leaves no place for the 'thick interface' of industrial livestock, slaughterhouses and processing plants, and might even be read as an attempt to nurture what Jacob Metcalf calls 'the intra-active, historical, material and discursive relatings that sustain ecologies of food'.[23] Metcalf calls for a more thoughtful approach to *in vitro* meat futures, arguing that 'a fundamental dilemma will be whether our moral obligations to reduce suffering (and other harms) necessarily leads to a world in which *organisms that can suffer are engineered out of it*.'[24] The pig in the garden contributes to efforts to 're-world technoscience', or 'leave the relations that sustain human and non-human life more livable'.[25]

The Critical Art Ensemble (CAE) shares Chin's and Next Nature's commitment to imagining humorous ways that high-tech foods might enter ordinary kitchens, but their cookbook maintains a dystopian view of the on-going dominance of food corporations and big brands. Published as a standalone artists' book and as an appendix in the book *Molecular Invasion*, the short cookbook *Betty Crocker 3000 Presents Food for a Hungry World* addresses 'those of us without access to the organic bounty and culinary luxury'.[26] Declaring that 'we have no choice but to embrace the genetically-engineered (GE) foods provided by the food industry', the 'recombinant dishes' include 'Monarch Wings Orientale', 'Mock-Lobster Recombinée', and 'Hormel Spaghetti Rings with Fishsticks' in the 'Kids' Korner'.[27] Most of the recipe ingredients include big food brands like Kraft and Tyson's. Ten years after the publication of *Betty Crocker 3000*, these brands continue to be embroiled in labelling controversies, recently over the word 'natural' appearing on products that contain genetically engineered ingredients.[28] *Betty Crocker 3000* celebrates the unnatural and the thick interface, drawing attention to how consumers are economically compelled to buy food from companies like Kraft and Pepsi.

Betty Crocker never addressed economic hardship or class consciousness. Brief mentions of 'cost' in her meal planning section include advice like 'buy less of the more expensive foods'.[29] Betty Crocker relies on the idea that brand names can become part of our daily lives by being personified as mascots who could be 'friends' with consumers. General Mills has kept Betty Crocker at work well beyond the lifespan of most 'live trademarks', including Mary Blake (Carnation), Mary Hale Martin (Libby's), Patricia Collier (Dole) and Ann Pillsbury (Pillsbury).[30] In her essay '"I guarantee": Betty Crocker and the woman in the kitchen', Laura Shapiro argues that these fictional home economists 'forged a crucial link between old habits and new foods'.[31] Live trademarks are technology evangelists, promoting new methods and products. Contemporary recipe artists adopt the cookbook and cooking show format to critique rather than evangelise. General Mills continues to promote big food through Betty's evangelism, dropping General Mills ingredients in her recipes just like CAE does with *Betty Crocker 3000*.

CAE includes price comparisons for organic versus GMO ingredients, and makes innovative use of prepared meals. For example, the recipe for 'Casta Casserole' recommends removing the pork chops from a Marie Callender's Frozen Dinner and repurposing them in a microwaved casserole. Foods emerge from packages rather than from animals or plants. Casta Casserole assumes that pork chops are best consumed after they have already been processed into a stockpile of frozen dinners. We understand ingredients to be components of packaged combinations, and the cook is tasked with deconstructing and recombining parts of pre-packaged wholes.

Culminating with the 'Future Feast GFP Banquet' and its centrepiece dish, 'Ragout Alba a la Provençale', *Betty Crocker 3000* traces the porous boundaries between grocery stores, laboratories and kitchens, suggesting that these spaces overlap more than we might have thought. Meat at the grocery store and on the table often results from laboratory research, even if animals do not move directly from laboratory to butcher (although they might move from laboratory to renderer). CAE's suggestion that 'scientist friends or molecular biology students' might be able to obtain GFP plants and animals for culinary purposes seems unlikely at first, but laboratory and culinary applications often commingle. Consider Pel-Freez, an Arkansas based company with two divisions: Pel-Freez Biologicals supplies laboratories with animal materials from rabbits, guinea pigs and other animals, while Pel-Freez Foods markets rabbit meat to consumers.[32] Laboratory animals, especially rabbits and dogs, find their way to new homes with humans through adoption agencies. Rabbits are more likely to be eaten than any other animal kept as a pet in the United States context. Seen as pets or meat, sometimes both at once, and also very common in laboratory settings, rabbits traverse laboratories, kitchens and markets. Cooking and eating a GFP rabbit addresses a neglected element of Eduardo Kac's *GFP Bunny*. Kac named an INRF GFP rabbit Alba and intended to take her home to live with him as a pet, ignoring the possibility that Alba would have been regarded as meat in many households. *Betty Crocker 3000* approaches Alba as a contraband edible, smuggled out of her laboratory home to grace blue-lit dining tables of cooks with contacts in biotech laboratories.

Choosing a cookbook format for this critique works on several levels. The narrative force of recipe instructions makes GFP meat and salad a speculative reality: as with Chin's cookbook, strange ingredients become real when we imagine them in our kitchens. At the same time, attention to costs and sources of ingredients heightens awareness of the complex food ecologies and financial choices we negotiate when preparing even the simplest meal. As with Greenpeace's *Good Food Gone Bad*, *Betty Crocker 3000* celebrates GMOs with a tongue-in-cheek humour designed to reveal the larger socio-economic systems at work behind familiar foods. Both *Betty Crocker 3000* and Chin's *Genetically Modified Foods Cook Book* use the cookbook format to suggest that these foods are already on the table or, in the case of *The In Vitro Meat Cook Book*, soon might be on the table. Even if we do not yet eat finger carrots and mouthy tomatoes, we nonetheless operate in a culinary space where the laboratory, the kitchen and the grocery store can never again be separated. Our cookbooks must adapt to these blurred boundaries and new technologies. What do genetic modifications taste like? Chin and CAE propose that these new technologies taste of glow, cannibalism and chimeric manipulation.

All three of these cookbooks use visual cues to extend the range of multisensory encounters we face when eating new technologies, and even extend the senses from taste to touch to more abstract understandings of how our bodies absorb and digest new technologies and how animals and humans share social space. *The Genetically Modified Foods*

Cook Book, *The In Vitro Meat Cookbook* and *Betty Crocker 3000* emphasise the social interaction of transspecies ingredients and encounters, as with Chin's mouthy tomato and earsprout pairing and Next Nature's pig in the garden, and social interactions outside of the kitchen, as with CAE's suggestion to ask 'biotech family' friends for GFP ingredients. These recipes interrogate such external systems alongside more intimate speculation regarding the body's capacity to absorb new foods, including GMOs. When eating organisms that traverse species and kingdoms we cross these boundaries ourselves. Chin, Next Nature and CAE suggest that we knowingly cross species boundaries with a sense of humour, always aware that eating technology catches us in unpredictable entanglements.

9

Carnal Light

with Eva Hayward

Ragout Alba à la Provençale

(GFP rabbits are being raised in select labs that will not release them into the wild but might sell them to enterprising cooks for a special banquet.)
1 GFP stewing rabbit cut into chunks (about 3 pounds)
(Marinate the rabbit in a mixture of wine, vinegar, olive oil, mashed garlic cloves, a bay leaf and pinch of thyme for 4 hours.)
2 cups canned chicken broth
2 cups each chopped onions, carrots, celery and red, ripe tomatoes
½ cup chopped GFP parsley

In an oven-proof casserole layer the chopped vegetables with pieces of marinated rabbit. Pour broth over the top. Cover tightly and bake in a medium hot oven for 3–4 hours. Arrange on a platter surrounded by green mashed potatoes. Sprinkle generously with GFP chopped parsley for maximum fluorescence.

(Cost estimate: GFP rabbit and other ingredients free to participating biotech families)[1]

~

The third verse of the Book of Genesis reads, 'And God said, Let there be light: and there was light'. Following the creation of earthly matter, light has illuminated Western stories from the beginning. Solar radiance travels through space and refracted planetary atmospheres to illuminate Earth's watery matter and from this encounter come lively forms. Life, as we are taught, has always been shaped by light even if sometimes inadvertently. And yet, meaty forms of light tend to be written from histories of Western thought in favour of an invisible medium that makes a knowable world visible: seeing light is a trope for making visible that which is invisible or perceiving things in a comprehensible form such that they are real. Jacques Derrida suggests that the language of philosophy is constituted by metaphors of light. Describing philosophy's reliance on light, he says that this founding relationship functions 'not only because it is a photological one – and in this respect the entire history of our philosophy is a photology, the name given to a history of, or a treatise on, light – but because it is a metaphor'.[2] As such, sunlight as a materialising wave metaphorises into transcendent light, a heavenly light promising liberation from the flesh.

The emergence of transgenic technologies has reintroduced light back to flesh. A central figure in this reintroduction is the jellyfish, *Aequorea victoria*, and particularly its green fluorescent protein (GFP).[3] Witness ANDI, the rhesus monkey whose stillborn twins continued to glow in death;[4] GFP fish made sterile for decorative purposes;[5] glowing kittens whose GFP expression marks the presence of a protein that resists feline immunodeficiency virus;[6] and transgenic pigs who glow as a means of tracing donor cells after stem cell therapy.[7] Illuminating biomechanics, bodily properties and chemical signatures, GFP expression and phosphorescence have been written into the bodies of many other organisms, including bacteria[8], yeast and other fungi, plant, fly and human cells. In 2008, Osamu Shimomura, Roger Y. Tsein and Martin Chalfie were awarded the Nobel Prize in Chemistry for their pioneering work in isolating GFP from *Aequorea victoria* in the 1960s, which produced the chimeric borderlands that GFP and transgenics inhabit today. Glowing plant and animal tissues pass through strictly visual pleasure into the articulation of carnal interdependence across species boundaries.

Drifting with *Aequorea victoria*, the host and source of green fluorescent protein, this chapter follows GFP as it moves on currents of bio-capitalism and animal industry. GFP drifts across kingdoms, phyla and species, blooming chimera and multiplicities that push against the perceptual and affective registers of laboratory walls, embodied selves, eating practices and living conditions. *Aequorea victoria* moves through laboratories, galleries, kitchens, medicine and visual technologies and into the bodies of multiple organisms. GFP-donor jellyfish are visualising but invisibilised forces, even as GFP-host animals steal headlines. Bioart projects such as Eduardo Kac's *GFP Bunny* (2009) are emergent forms of human kinship arising from biotechnologies. Such creatures push against the perceptual and sensorial elements of laboratory walls, embodied containers, eating practices and living conditions. GFP hosts and donors reveal the carnality and radiance of naturecultures – sites of social and biological constitutive formations – by welcoming us to a messy, multidisciplinary table where light is being eaten.

Somalumenals, tranimals

Visuality as corporeality, light as meat: somalumenal. Soma and somatic are derived from the Greek *soma*, meaning body. For mammals, somatic cells make up bone, skin, blood and other tissue. However, they are not involved in sexual reproduction for organisms that reproduce sexually. *Lumen*, Latin for light: light, lantern, lamp and also clarity and understanding. Lumen also lights the day as a unit of luminous flux, the amount of visible light emitted from a source. But Lumen is also 'the central cavity of a hollow structure in an organism or cell': in nematodes, an esophagus; in fungi, the space bounded by tissue.

The neologism 'somalumenal' registers the multiple etymologies of these words. How might luminosity, effulgence and illumination be about embodiment or corporeality? Light is transubstantiated into flesh and matter, rather than the body transcended by the thrills of enlightenment and radiance. Somalumenal has particular resonance for transgenic organisms that have been designed to glow. Through genetic transfers, fish, monkeys, rabbits, mice, pigs, cats, bacteria and plants have been modified to express kinds of glow. Using luminescent jellyfish genes that code for green fluorescent protein, geneticists have combined the jellyfish protein with other organisms, producing glow in uncanny sites.[9]

An enmeshment of 'trans' and 'animals', tranimals is a synecdochically imagined correspondence written as a portmanteau word (a blending of sound and meaning).[10] Hybrid forms and liminal states of tranimals represent subtle or even explosive changes in our understanding of bodily transformation. These are instances within which we can see more closely the appearing and disappearing boundaries between the human, the post-animal (human and non-human), the in/un/human and the animal. This is not necessarily about cross-species identification, but is rather about a somatic and sensual synthesis that manifests synecdochically[11] rather than metaphorically.

Tranimals also plays with the prefix 'trans' as a plurality – as in animals – to suggest the on-going labour of cross-species agents. Trans suggests the energetic and material crossings that disrupt bifurcated categories. Animals suggest literal organisms, not metaphoric or purely representational beings that exist only in anthropocentric intentionality.

Following Donna Haraway's metaplasm, we inflect tranimals with its tropic kin of antimeria that turns nouns into verbs, that mobilises, incites, activates persons, animals, places, things and ideas.[12] From Greek: *anti-* 'instead of' and *mereia* 'a part': incitement rather than determination, for example, 'I am rubied by your attention'. So it is not just the encounter that matters but more exactly the relay of energies and forces that constitute ontology. Ontology is about what there is and what debts we owe to it. We exist because our bodies must respond to life's instances so as to sustain our vitality. We must respond to these insistences through, we would argue, that which we call sensation. The corporeal 'prehensions' of metaplasm are always immersed within a sensuality that generates a 'critter' whose sensibility manifests with its emergence. Responsiveness is the condition of emergence. So, where metaplasm invokes the remodelling of flesh, antimeria is the transmission of beingness through intensification. In this way, nouns are transitional responses to relationships rather than direct products of environments. Antimeria expresses the sensual charge of encounter, its amplification, allowing for the percussive distribution of affect. Remember that the transduction of force, the passing through or between creates remainders of uncircumscribable passings that result in the *feel* of beingness. Boundaries remain refracted interfaces of passage, prepositional orientations. Sensuous being is the unmetabolisable *more*, the residue of passing, of sensations moving across bodies and objects that are felt by the mediation of the sensorium. A jellyfish stings human skin, leaving an empurpled mark that inscribes the boundaries of tentacle and skin (somatic technologies), but the excitation (sensorial transit) of burning and blistering energy remains partly unmetabolisable – this remainder is differentiating and binding. So excitation is not necessarily predicated on positive joins of ever expanding units of life, but emerges between them and therefore can be generated through haemorrhaging, excising, amputating.

Alba

The playful annunciation of tranimals describes the authorial intent of transgenic artists such as Eduardo Kac. A fiercely serious art project, *GFP Bunny* also invents and explores boundaries shaped by questions of consequence and responsibility. What debt does Kac owe Alba, the *GFP Bunny*?

'Alba' means glowing white, a hybrid rose and also a song sung to meet the dawn, inviting us to imagine breaking fast with glowing light. Alba eats, sleeps and interacts with human others as any rabbit might, even if at the same time she carries a new line of glowing, transgenic mammal-jellyfish genes. Corporeal kinds of luminosity, refulgency and fluorescence define Alba's status as a 'tranimal', a hydrozoan 'becoming-with' mammalian tissue; that is, an organism animated by trans-formations of life. She is literally rabbit, jellyfish, human and technology.

Transfection releases jellyfish materiality into Alba's bodily frame, making glow the expression of another outside brought inside. Tranimals, translife and transgenics of all persuasions emerge as manifestations of error or drift. Alba further refines trans-practices by attending to questions about light, enlightenment and embodiment. Alba's flesh becomes light; her glow is refracted through tissue, making bodies materially luminal.[13] In this way, light is no longer the extra-terrestrial force of more heavenly bodies, but effulgence bound to meatier substrates. This carnal light offers a supplemental reading of visuality not rooted in reflection, refraction and diffraction of solar light to illuminate discourse.

Aequorea victoria is the source of Alba's green glow. Found along the coast of the Pacific US, they are composed of more than 90 per cent water. Most of their umbrella mass is a gelatinous material called mesoglea. They do not have a central brain, but rather a loose network of nerves called a 'nerve net'. With light-sensitive organs they do not form images but detect light. At the edge of their bell, 150 uneven tentacles appear. The larger jellies possess gonads for sexual reproduction, which run most of the length of the radial canals. Muscular velum ring the margin of the bell, which aids in locomotion through muscular contractions of the bell. Large adults are frequently found with symbiotic amphipods attached to the subumbrella or even occasionally living inside the gut or radial canals. Alternating between asexual benthic polyps and sexual planktonic medusae in a seasonal pattern, these jellies have a dimorphic life history. Voracious predators in their own ecosystems, they gorge on other jellies, copepods and invertebrate larvae, even eating one another. Ensnaring with long tentacles packed with stinging cells, they ingest through a highly contractile mouth.

They are pelagic drifters. They drift, are literally adrift. Drift is also tropic. Drift: to be carried slowly by a current of air or water; move passively, aimlessly or involuntarily into a certain situation or condition; digress or stray to another subject. For some matter, drift also gathers, heaping with out intentionality, without consciousness. In noun form: a steady movement or development from one thing toward another; the general intention or meaning of an argument.[14] To drift enacts movement, changing something into another form or to transfer to a different place or context. Drifts are poetic joins in which there is some drifting together, some mixing of ontologies into larger organisations. For a jellyfish, its body is index of its environment and glow is the bodily response to tactile vibrations. When provoked, they give off a green glow. Through a release of calcium in the body, which interacts with the photo-protein of aequorin, the jellyfish flashes blue. GFP transduces the blue chemiluminescence of aequorin into green fluorescent light. While jellies drift through their environments, glowing in response to provocation, their capacity to glow separates from their bodies to be sent adrift as mutation, as error in codes of life.

What can we say about the transfer of luminescence from jellyfish capacities and lifeways into the flesh of other species? How does this blend of tissue – invertebrate and vertebrate – re-order not only life but also light? Writing of Jakob Von Uexküll, Dorion Sagan reflects that 'life is not just about matter and how it interacts with itself, but how that matter interacts with interconnected systems that include organisms in their distinct perceiving worlds'.[15] Uexküll's bridging of phenomenology and biology helps us feel our ways into the umwelts (perceptual bubbles) of the most unidentifiable animals (ticks, spiders and other inverts) teaching us that organs of animals are counterpoints of lived environments. The body of the animal is the inverted map of its environment but mediated by its capacity to interpret habitat through bodily sensations. Jellyfish and rabbit play out a discordant tonality (Uexküll uses music to define the laws of nature – life is musical response, at least a duet). They are not companions, not shaped by a contrapuntal composition, but by the effects of animal capitalism.[16] They are fellow travellers through stages of evolution of the consumer product and in that they share cadence, becoming a new harmonics or improvisation.

In this way, sensuousness is at work in somalumenals. Through the philosophical tools of phenomenology Alphonso Lingis explores somatic sociability and its sensuous rapport. Sensation is more than stimulation of our sensory or nervous systems. It is an intensity that pressures us to feel our cohabitation in a sensible environment that opens us outward. Lingis writes: 'level of sensation would be the original locus of openness upon things or contact with them'.[17] In this way, the subject is 'involuted' into the sensuousness of being; the substance of self is generated from the excitations of affect and desire within a sensuous environment. The sensorium receives the movements of the world not because it is pure receptivity, but because the limits of sense shape it. A sphere where the sensual resonates and territorialises itself, it percusses to the expressiveness of beings, things, events through the mediation of its own flesh. The moreness of the sensible world presses on our sensibility, revealing that liveliness is responsiveness, to feel folded into the plenitude of the world, which is always defined by the refrain of one's own materiality. 'To sense is to sense the substantial', which is the foundation of our subjectivity.[18]

Alba is not a singularity, not a pre-existing solitary species. She foregrounds the labour of multiple species as parts constituting a collective being. Moving beyond now familiar forms of seductive hybridity, tranimals invoke the social, affective and sensorial response-abilities of becoming and eating. Alba is a frame, a distinctly rabbit-like frame, in which the expressiveness of GFP is unleashed, in which GFP's luminous excess becomes more than a jellyfish's response to environmental cues. In other words, as a bioartwork, Alba is more about intensity than aesthetics and equally about sensation as *techné*.[19]

We share our subjugation to sensation with other sensuous beings, beings that respond to ecosystems through their own sensoriums that define their own milieus. An individual sensorium is nothing other than an improvisation of others synchronising with their own environments. For Lingis, sensations are amalgamations, made of movements and affects that shape the energetic relay between subjects and objects. Affects are about becoming or emergence, while percepts originating in environmental harmonisation are also the excess of this evolutionary force. Sensations reside between sensibility and sensuousness – they are sensate sense or equally sensible sensuality, between subjects and objects, in the encounter, at the point where they shape the other. Consequently, sensation distributes affects and percepts across bodies. The body restrains affection and perception, but excitation unleashes bodily organisation at the site of contact. Excitation, then, is the zone not only of contact but an expressive threshold, a vital interface of emergence. Vital energies describe the conditions of our on-going origination, our future selves.

The environment that is webbed into a jellyfish's capacity to illuminate is made present in the rabbit, but creates an entirely different register of reactivity. Alba is a becoming of multispecies materialities and energies, an artwork, to borrow from Elizabeth Grosz 'in which life folds over itself to embrace its contact with materiality'.[20] Alba is a composite; her rabbit form mammalises light. With the help of jellyfish kin, she locates illumination in skin, ears, eyes, in her organs. Even as our human eyes are drawn to Alba's fluorescence, her sensitive whiskers and glowing ears augment her weak but now bio-luminescent eyesight. Whether or not she herself registers the glow, for her primate caretakers and audiences, the glow intensifies crossed boundaries. That is to say, in her flesh, Alba is tranimals; to us, her glow expresses, sensitises and excites that fact. Her glow is a dynamising force, making her matter more. Although Alba cannot choose when she glows, jellyfish substances remain activated, active, responsive to stimulus, excitation. Alba is a genetic improvisation with bio-capacities syncopated. Through her glow, her artfulness, 'we are reminded that species are not just relationships, nor are species ever directly in relationship with other species, we are resonances and dissonances of intensification, energetic cadences of one another' – a sensorial ensemble.[21]

Alba in laboratories, kitchens, as food

All GFP tranimals travel between contexts and spaces. Alba's movements unravel categories of companionship and reveal vulnerable overlaps between previously autonomous systems. As sites of human animal labours of science, of eating and of serving and dining, laboratories, kitchens and dining rooms emerge as generative, troubled environments.

Alba was born and made in an 'art lab'. This is not an ordinary experimental laboratory, but rather a site where science and art are compounded. Art laboratories come into being in a variety of ways. They may be ephemeral, the result of a temporary recontextualisation of activity in an industry or government laboratory. They may exist outside of institutional confines, in a kitchen, a shed in the back yard or in the home studio. In its rarest incarnation, the art laboratory might be established with full institutional support, with staff and resources dedicated to art–science amalgamations.[22] Alba was born in a repurposed government laboratory, France's Institut National de la Recherche Agronomique (INRA). Alba's birth, presence and death remade the INRA as an art space.

Alba's laboratory home was both the site of Eduardo Kac's art intervention and shelter to innumerable laboratory rabbit kin. Upon her death in 2002, at four years old, genetic researcher Louis-Marie Houdebine reflected on Alba as one of many: 'I was informed one day that bunny was dead without any reason [...] So, rabbits die often. It was about 4 years old, which is a normal lifespan in our facilities.'[23] Alba's real and imagined homes have been shaped by the somatic sociality of an art bound not by representation but by the uneasy space between scientific experimentation and domestic cohabitation. As with rats, rabbits are brought into the home as pets; even as vast colonies their more carefully bred kin occupy laboratory spaces like the INRA and breeding facilities like Renova Life Sciences.[24]

An emphasis on 'a social and interactive context' and 'public respect and appreciation for the emotional and cognitive life of transgenic animals'[25] allows Kac's art laboratories to traverse domestic and public zones, critically commenting on overlaps between laboratory production of materials brought into intimate domestic spaces, such as food and medicine. Alba's social intervention reminds us that bioart production is the result of art–science convergences not wholly attributable to advances in biotechnologies and genomics. Suzanne Anker writes: 'the chimera must be understood in a social context: These beasts have always reflected the social, scientific and religious circumstances of their time'.[26] Although tranimals are workers, not mythological, not fabulous but pragmatic and instrumental, they reflect and mirror the material circumstances of our time in much the same way as do Anker's chimeras. In this overlapping zone of science and art, Alba comes together as a being, a composition and a 'biocultural' experiment.

While bound by GFP to jellyfish and the sea, Alba does not swim well and instead hops across laboratory and domestic spheres.[27] Laboratory rabbits are historically connected to toxicology and reproductive testing and valued as living incubators for antibodies and antiserums. Renova Life sells GFP rabbits for researching 'human related diseases, such as CVD, HIV, regenerative disease and cancer'.[28] Renova warns that the rabbits are 'not for household use', although their home page incorporates Alba's image into a life sciences montage. Alba was named with the goal of leaving the laboratory to be adopted by Kac's household in Chicago, where she would serve none of these laboratory functions, instead existing as pet and as somalumenal curiousity. Kac was unable to bring Alba home, but other laboratory rabbits do find homes outside their labs of origin. The Rabbit Welfare Association and Fund encourages the adoption of laboratory rabbits, declaring that 'former laboratory rabbits are sometimes available for adoption (all legal and above board!) and such animals are generally extremely well handled and always absolutely healthy'.[29]

The technologies responsible for Alba's somalumenesence have been widely employed for some time and can no longer be thought of as artful innovations.[30] When Kac refers to Alba's 'formal and genetic uniqueness',[31] we appreciate her unique genetic qualities by way of imagined domestic encounters that become transgressive when laboratory animals are involved.[32] What does it mean to live with, eat with and companion tranimals? What does it mean for pets to live in research laboratories? The spectre of Alba-as-pet, moving beneath a blue light in the living room of Kac's home, evokes the invisible labour of jellyfish, research on the proliferation and patterns of cancer cells and the sensation of somalumenality. Encounter trumps Alba's technical becoming as we are asked to imagine meeting Alba outside of her laboratory home or meeting her differently within the walls of an imaginary art laboratory mapped onto the INRA.

The idea that a laboratory rabbit would live in Kac's household as a companion animal suggests that just as art has remade the laboratory, artful tranimals might remake the home, especially the kitchen and dining room. Alba straddles slippery pet/food distinctions: many humans cannot read the rabbit as pet without also regarding her as food. Often interpreted as a site of human exceptionalism, cooking must be reconfigured in the rabbit and ruminant context. In the Critical Art Ensemble's short cookbook 'Betty Crocker 3000 presents food for a hungry world', the recipe 'Ragout Alba à la Provençale' cooks and eats Alba and her laboratory rabbit kin.[33]

While GFP tissue continues to glow after death,[34] the suggested 3–4 hours in the oven for this ragout might diminish GFP's capabilities. Replacing dining room light fixtures with blue bulbs should provide a necessary boost. Here, eating Alba is not a science fiction future, but a present-day exercise in the ethics of manipulating animal flesh for experiment, for the table or both at once. Even though GFP rabbit meat is not available in supermarkets, the utility of GFP in the design of transgenic flora and fauna means that the plants, fish and meats we eat today might already contain GFP. Human ingestion of GFP signals another type of drift distinct from the allelic drift of the jellyfish. Glow drifts through the bodies of consumers.

The stakes of eating tranimals and allowing for this process by virtue of being a 'participatory biotech family' are both fantastic and uncanny when we recall the transspecies sequence living in Alba's tissues. Ingesting, killing and eating jellyfish usually take the form of dried crisps sprinkled into salads and Chinese food. (*Aequorea victoria*, the Pacific jellyfish source of the first GFP derivations, is not considered edible.) Ingesting jellyfish by way of tranimals like Alba implies a more complicated food relationship between humans and jellies, where jellyfish genes drift through the bodies of animals whom human animals enjoy as companions and as food (rabbits being an example of the slipperiness of pet/food distinctions). Donna Haraway reminds us that:

> *Companion* comes from the Latin, *cum panis*, with bread. Mess mates at table are companions. Comrades are political companions. Business and commercial associates form a company, which also connotes a guest, a Medieval trade guild, a fleet of merchant ships, a military unit and the Central Intelligence Agency. As a verb, 'to companion' is to consort, to keep company, with sexual and generative connotations always ready to erupt.[35]

What happens when messmates at the table become messmates on the table? Human animals eat with rabbits and eat rabbits. The human animal's decision to accompany another from a shared table into the stomach requires not just slaughter, preparation and cooking, but also an ability to ingest one's friends.

As described by Robert Foley on the radio programme 'The Naked Scientists', the human animal's cooking behaviour is 'an extension of things we see in other animals'.[36] Alba's first digestion, which produces edible caecetroph 'night droppings', performs many of the same functions as the oven or stove. Living with rabbits means cooking in parallel, cooking for or being cooked by another. Julie Ann Smith writes about how faecal matter becomes food not just for private consumption, but also for human-assisted care of other rabbits:

> Rabbits pass their food twice, the first time consuming their feces directly [...] we nurse rabbits with digestive problems by inoculating them with gut bacteria obtained from the feces of healthy rabbits. We prepare a 'fecal cocktail', which is syringe-fed to the sick rabbit.[37]

Eating with rabbits requires the sharing not just of bread but of intimate gut bacteria. This exchange of gut bacteria is multidirectional, but also maps a power dynamic where human gut bacteria only enter the equation after rabbits die and bind the human animal to the rabbit they have been accompanying through taste and digestion.

As rabbits cook their food, so they are also sometimes cooked as meat by their caretakers. The fictional kitchen where readers prepare 'Ragout

Alba à la Provençale' references rabbit meals made at home that are less luminous, featuring rabbits from backyard hutches and commercial rabbit farms. Although rabbits are often raised commercially for slaughter, because they are small and easy to keep they are also a popular backyard animal and a popular indoor pet. Rabbit meat has been recently labelled a 'crisis meat' in the US, describing a turn towards rabbit during lean times.[38] Rabbits move between living room, backyard and oven with dramatically more frequency than pets like dogs, cats, rodents and birds. Stories of rabbits crossing contexts abound: a large albino rabbit intended for a fraternity house dinner was rescued from a cooler and now lives in cooperative housing in Santa Cruz, California. Two rabbits kept as pets were sold to and eaten by the families of elementary school classmates in Vermont. Rabbits walk the line between livestock and pet with a household mandate that 'We can have rabbits, as long as some of them are for the freezer'.[39]

Writing about her experience fostering house rabbits for adoption, Smith describes how her home has been undone and remade by cohabitation with rabbits, reminding us again that ontology is a practice of co-constitutive awareness and responsibility.

> In my own house, rabbit-proofing meant that most of the furniture was made of metal, electrical cords were fastened behind furniture or covered in hard plastic or metal tubing and protective wood strips were tacked on to wood baseboards and wood trim around closets and windows [...] So-called 'litterbox training' primarily meant capitalizing on the rabbit habit of urinating consistently in one or two places. We simply put litterboxes where the rabbits decided to eliminate. Many of us found it easier to change ourselves than the premises. At present, the rabbit who lives in the bedroom is excavating my mattress. She bounces around inside the dust cover and chews the wooden frame around the metal springs [...] Frankly, I love this way of living, this version of 'becoming animal'.[40]

Presumably something similar would have happened in Kac's home in Chicago. For Kac, changing the self and adapting to cohabitation with rabbits would have had wider consequences, invoking Smith's engagement with what Haraway would call 'becoming with' rabbits: 'the play of companion species learning to pay attention'.[41] Smith remakes 'training' into 'paying attention', reconfiguring her home as a place where she can 're-imagine the rabbit'. Imagination overrides other companion species possibilities in Alba and Kac's situation.

Complete with models for how humans should leverage social and employment contacts to procure GFP meat, 'Ragout Alba à la Provençale' addresses both people who eat pets and people who eat laboratory animals, questioning how laboratory animals, pets and meat are related. Certain tranimals compel a meaty treatment, even in a laboratory context. For example, a 2006 article about GFP pigs born in Taiwan is titled 'Is this sausage supposed to be green?',[42] implying that GFP pigs herald the supermarket availability of bioluminescent bacon. Viewing the bodies of pig tranimals as bacon becomes a prerequisite for understanding the science behind their births and the research applications for which they have been engineered.

Eating light

With CAE's recipe we can imagine that glowing food intensifies vision through tactile registers; the verb 'to taste' is conjugated through sight. Glowing at the edges of the mouth, 'Ragout Alba a la Provençale' cannot be digested without visual recourse. For *homo sapiens* consuming this entrée, the mouth is solicited by light, undoing the eye's propriety of vision.

Following Derrida, Cary Wolfe conceptualises Kac's use of GFP as a 'feint' or 'lure' that whets our 'carnophallogocentric visual appetite', yet, importantly, doesn't fill our stomachs. For Wolfe, Alba's fluorescence yields not the invisible or the hidden but a posthumanist repositioning of visuality itself: 'Kac *situate*[*s*] the visual in ways that fundamentally trouble how we have typically indexed the (human) animal sensorium to the human/animal ontological divide.'[43] Alba's GFP expression flirts with but denies humanist ways of organising visual information. As she appears in 'Ragout Alba a la Provençale', Alba would force our human 'carnophallogocentric visual appetite' down our throats as food, growing light receptors in our taste buds. Although for Alba this is an unbecoming, a transformation from companion to meat, for the humans involved, vision is reworked by orality, by taste and jellyfish bio-materialities are sent further adrift.

Wolfe's conceptualisation of the 'feint' and 'lure' as posthumanist shifts applies to popular discourse about the science underpinning Alba's existence as much as to her life in art. GFP acts as a marker, a way of tracing biomaterial transformations in tissues. Like knockout mice, GFP tranimals host jellyfish genes in order to make other genetic modifications visible. For example, Renova Life offers comprehensive services for working with GFP rabbits, whose jellyfish genes frame a range of questions:

> The new lines of GFP rabbits can be used for study many human related diseases, such as CVD, HIV, regenerative disease and cancer. The specific relevant or derived GFP rabbit cell types can be used for gene therapy in animal models. GFP transgenic rabbits can also be used for deriving embryonic stem cells (from the early embryos), induced pluripotent stem (iPS) cells and further differentiation into many specific cell types. Those cells carry the GFP gene and can be traceable during the development and differentiation, as well as cell migration during the process, especially during the study of cell-based therapy.[44]

Tranimals labour to 'carry' GFP. Renova Life reminds us that 'quality control makes sure GFP gene is transmitted in all cell type, the animals and derivative tissues/cells express GFP'. Alba was not engineered to glow any more than a pig was engineered to become a sausage. The tired narrative of the ghostly sausage haunting a living pig contrasts with the tidal drift and shift of 'jellyfishification'. Writing in *Science* magazine, Erik Stokstad considers environmental risks of GM farmed fish, weighing the benefits of relieving native populations by consuming genetically engineered sterile superfish against the possible escape and contamination of native fish populations by the same tranimals. Stokstad investigates the material conditions of sterilisation procedures, imagining that scaling up laboratory sterilisation processes to industrial levels would inevitably produce a margin of fertile error.[45] Somalumenal tranimals host a posthuman potentiality inside their every cell.

Having followed GFP from jellyfish mesoglea to rabbit skin to human stomachs, we find ourselves bound within Francis Bartkowski's empathic, kissing-killing knot. Her attention to the kiss marks, the orality of cross-species intimacy as 'the threshold of destiny and desire [...] a space of permeability'.[46] Ingesting and inserting GFP signals another modality of getting under the skin, marking skin, kin and species not as discrete entities but as spaces of sensuous interchange. GFP generates networks, zones of cohesion and carnal assemblages through sensation, expressiveness and intensification. Adrift with tranimals, we too are beginning to become jellyfish, further sensationalising the adage 'you are what you eat'.[47]

10

From Sanitation to Bioremediation

Smog Tasting Meringue Recipe

Ingredients

Makes 16 meringues
4 Egg Whites, room temperature
200g Caster Sugar
Lemon Slice
Pinch of Salt

Steps

Takes 2 hours total

1. Preheat the oven to 540° and line 2 baking sheets with parchment paper.
2. Make sure all cooking equipment is very clean. Wipe a large mixing bowl (ideally not plastic) and your electric mixer beaters with a lemon slice to help stabilize the foam.
3. Place 4 egg whites in the mixing bowl with a pinch of salt.
4. Beat egg whites on medium speed until the mixture stands in stiff peaks.

5. Slowly add the caster sugar one spoonful at a time, beating on high. Don't over-beat. The mixture will be thick and glossy when ready.
6. Fold in ½ desired smog spice mix (see recipes for each city).
7. Heap a tablespoon with the mixture and scrape onto baking sheets, creating about 16 small 'clouds'.
8. Sprinkle remainder of spice mix across the tops of the meringues.
9. Bake until crisp, 1 ¼ hour to 1 ¾ hour. Let cool on the tray.[1]

~

Ecology and the home have always been linked. Influenced by Ernst Haeckel's *oekologie*, from the Greek *oikos*, meaning house, Ellen Swallow Richards (1842–1911) was among the first scientists to make the link explicit and relevant to the emerging field of Home Economics. Her work in 'sanitary chemistry', or later simply 'sanitation', prefigured public health and ecology.[2] Richards's commitment to modelling and visualisation and her efforts to engage multiple publics mark her research as foundational for collaborative art-science approaches to bioremediation, despite the troubling ideological premises underpinning the research of many nineteenth-century scientists invested in sanitation.[3] Inspired by this ecological reading of sanitation as an early example of what visualising and intervening in anthropocenic landscapes might look like, this chapter takes a long view of bioart history by thinking taste and agriculture together with environmental remediation.

In the late nineteenth century, sanitation was an interdisciplinary field that included studies of industrialisation, public health and domestic interiors. By gathering data about pollution, sanitation research supported the first efforts to implement environmental legislation. Although arts organisations and artists were not directly involved in these early efforts, responses to industrial pollution animated many of the artworks of the time, indicating the cultural importance of this research.[4] Outsiders led sanitation projects – the women involved in the Home Economics movement were not policymakers and were often not taken seriously as scientists. This outsider and amateur status flips from a hindrance to an advantage for artists working today – pursuing bioremediation as art has allowed for flexibility and unexpected outcomes that would have been impossible were the same projects pursued in strictly science or policy contexts. Sanitation provides a context for contemporary bioremediation artworks, and artists emerge as critical interlocutors for reclamation efforts.

Although both projects were created before bioart cohered as a set of practices, Helen Mayer Harrison and Newton Harrison's *Survival Piece No. 2: Notations on the Ecosystems of the Western Salt Works (with the Inclusion of Brine Shrimp)* (1971) and Mel Chin's *Revival Field* (1991–ongoing) predict bioart's interest in living systems, or in Robert Mitchell's terms, 'the vitality of media'.[5] More recently, Natalie Jeremijenko's *Cross(X)Species Adventure Club* (2010–ongoing) focuses on how bioremediation might be enacted through food webs, while the Center for Genomic Gastronomy's *Smog Tasting* (2012) measures pollutants with meringues, engaging taste in our understanding of air pollution. Together, these projects model digestive approaches to environmental remediation and visualisation. Just as Richards enfolds the body in the home and the home in the ecosystem, these works facilitate rich parallel thinking that maps body to earth while including the labour of non- and posthuman actors.

Helen Mayer Harrison and Newton Harrison's *Survival Piece No. 2: Notations on the Ecosystems of the Western Salt Works (with the Inclusion of Brine Shrimp)* (1971) yielded salt, fish and 'three-dimensional color-field paintings' as brine shrimp consumed algae in saltwater ponds of various salinities.[6] The salt ponds housed complete ecosystems, with each pond's changing salt levels producing visible and edible changes. Increasing salinity compels *Dunaliella salina* algae to produce carotenoids that colour the pools with deep orange and red hues. Departing from contemporary earthworks that 'used earth as material', the Harrisons's early projects were 'among the first to deal with ecology in the full sense of the term'.[7] Consider the difference between large-scale earth-as-material works like Michael Heizer's removal of 240,000 tons of Nevada desert for *Double Negative* (1969–70) and the Harrison's *Making Earth* (1970), a composting project that later evolved into an ambitious park reclamation. In contrast to earth artists 'who worked with earth as if it were an inert substance, [...] *Making Earth* was an intimate investigation into elements that maintain the life web'.[8] Like *Making Earth, Notations* manipulated living systems, was time-based, and meditated on species survival.

The *Notations* ponds were situated outside the 1971 exhibition of the Los Angeles County Museum of Art's 'Art and Technology' programme, which paired artists with technology companies. The Harrisons showed work both inside and outside the gallery. Inside, Newton Harrison created *Encapsulated Aurora* in collaboration with Jet Propulsion Laboratory. *Encapsulated Aurora* was an elaborate and technologically sophisticated sculpture that captured and manipulated glow discharge. Glow discharge forms by passing electrical current through low-pressure gas (as with fluorescent and neon lights). *Encapsulated Aurora* houses mutable plasmas in custom-built tubes designed to respond to participants moving through the installation (and consequently altering the electrical currents passing through the gas captured by the tubes).[9] Both *Encapsulated Aurora* and *Notations* were concerned with colour fields, containment and chemical processes, but the indoor work relied on electricity to power delicate vacuum chambers, while the outdoor work powered itself with solar energy and hardy living systems.

'Notations' also produced tangible outcomes. It was a farm, factory and kitchen at once. Both shrimp and salt were harvested from the ponds; if the project were to be 'scaled up, at maximum production, the piece produces about 18,000 pounds wet weight an acre'.[10] Enacting a future-present collapse of aesthetics and survival, *Notations* asks, how might we sculpt the earth – not as inert material, but as a complex, living system? Craig Adcock argues that 'the art of Helen and Newton Harrison reminds us that our primary challenge as a species may boil down to making aesthetic choices'.[11] These aesthetic choices dovetail with many of the research questions that animated ecology as a discipline and one of its applied manifestations, sanitation and Home Economics. As a species, what kinds of environments should humans cultivate in order to feed a large population, transform one material into another, and positively impact the planet?

Questions of survival and environmental impact helped catalyse ecology as a discipline. The 1893 Chicago World's Fair imagined the future by presenting an ideal city filled with progressive exhibitions, from Eadweard Muybridge's purpose-built Zoopraxographical Hall to a Westinghouse-sponsored demonstration of Nikola Tesla's electrical inventions. Ellen Richards created the Rumford Kitchen for the fair. Humble and simple when contrasted with the spectacles staged by

Muybridge and Tesla, the kitchen served lunch, coffee and tea to 10,000 people over the two months of the exhibition. Although at first glance these projects could not seem more different, the kitchen aspired to some of the same outcomes as *Notations*. In her report on the project, Richards explains that the Rumford Kitchen was dedicated to 'illustrat[ing] the present state of knowledge in regard to the composition of materials for human food'.[12] What do people need to survive and thrive? How might we prepare food more efficiently? The kitchen and the ponds share a commitment to scaling food production and consumption, and modelling ways of living and eating differently. Survival in 1893 looks very different from survival in 1970, but the need for models, storytelling and speculative design persists.

Notations was foundational work for the Harrisons. The installation 'marked the beginning of their decades-long collaboration and their decision to work exclusively with projects that advantaged the life web'.[13] *Notations* and its successor *Portable Fish Farm: Survival Piece #3* (1971) were among the last of their works to be contained by gallery walls. After the *Survival Pieces* the Harrisons started to employ documentation and models to tell larger stories that 'reverberate at least on continental levels'.[14] *Notations* was also a foundational work in the history of bioart in both practical and theoretical ways. The Harrisons's collaboration with Dr Richard Eppley of the Scripps Institute of Oceanography predicted the steady rise in art-science collaborations that has created demand for spaces and organisations like SymbioticA and Arts Catalyst. Works like the Harrisons's *Survival Pieces* also make a theoretical intervention by providing a historical context for bioart projects that privilege food, eating and systemic approaches to ecological changes. The *Survival Pieces* are bioart projects that predate the wave of digital media and genetic technologies often cited as origin points in bioart history.

Moving forward 20 years to the early 1990s, 'deal[ing] with ecology in the full sense of the term' has occupied a wide swath of contemporary art practitioners.[15] The Harrisons's pioneering and ongoing work with living systems keeps company with an ever-increasing number of environmental artists and designers working in every medium to produce theoretical and practical work that addresses sustainability, ecological balance and the concept of the anthropocene. Bioremediation becomes increasingly important to environmental artists, with Mel Chin's *Revival Field* proving to be an enduring and iconic example of land reclamation-as-art. Collaborations with scientists remain fundamental to the interdisciplinary set of practices that started with the Harrisons and became bioart. As the project developed, *Revival Field* depended on collaboration with scientists to design both concepts and protocols, but at critical moments also depended on its status as art to attain funding and exist in the world. Addressing the National Endowment for the Arts in an effort to convince the agency to fund the project as art rather than dismiss it as strictly science, Chin uses the material language of art making to explain that *Revival Field* aims 'to sculpt a site's ecology [...] the material will be toxic earth and the tools will be a scientific process utilising heavy metal leaching plants called "hyperaccumulators" '.[16] Echoing Adcock's assertion that 'our primary challenge as a species may boil down to making aesthetic choices', *Revival Field* takes sculpture as a conceptual framework for shaping the earth.[17]

Revival Field began during a point in Chin's career when despite critical success with art objects, he retreated from object making, determined to subject himself to 'a conceptual fire'.[18] After encountering Terence McKenna's theory that jimson weed might be an effective phytoremediation technology, Chin reached out first to McKenna then to a wider circle of scientists and botanists in search of plants capable of extracting metal from soil (jimson weed, a psychedelic, turned out to be a poor choice for land that supported cattle or other grazing animals). Having contacted several scientists working in the emergent field of environmental remediation, Chin began work on *Revival Field* after talking with and reading papers written by agronomist Rufus Chaney. *Revival Field* seeks out sites of pollution and seeds them with 'hyperaccumulator' plants that absorb heavy metals in a quantity such that the stalks and roots of the plants could yield ore.[19] The first *Revival Field* was a superfund site in Minnesota, and there have been several additional sites since then.

Art and science combine in *Revival Field* to produce the first replicated field test of Chaney's laboratory work. The public benefit and scientific import of the project matter to Chin, as does its long duration: the project will outlive all of its founding collaborators. Of the project, Chaney remarks,

> we wanted to illustrate with art a possible solution that would help citizens [...] Mel and I agree that *Revival Field* is not done until it has become a technology [...] It will be done when it's part of the normal way society solves problems.[20]

Their desire for societal and environmental transformation through a shift in technology recalls the efforts of early public health advocates, especially those who, like Ellen Richards, looked to the water supply as a critical area for remediation. Identifying pollutants is only the first step. After becoming aware of contamination and mapping its presence, technologies of remediation have a chance to become routine. Phytoremediation may seem like an unlikely candidate for becoming a routine technology, but recall that before Ellen Richards's Normal Chlorine Map, watershed management and water treatment were unheard of, while today water treatment technologies are so ubiquitous as to be invisible. Mapping can itself be very challenging. One of the ongoing obstacles in the project was securing a polluted site whose owners acknowledged its contamination. By agreeing to host the project, site owners, often corporations, would admit to having polluted the site in the first place – and not many owner corporations were willing to own up to neglecting the land they contaminated.

Revival Field proposes a different way of managing the environment, a form of 'sustainability otherwise' akin to Zylinska's 'bioethics otherwise'. The hyperaccumulator plants act against conservationist desires to protect the environment by isolating and containing contaminated disaster zones and pristine 'wildernesses' alike, insisting on transformation and active intervention rather than protection and conservation. Responding to Donna Haraway's call for a cyborg politics that 'advocates pollution', phytoremediation technologies embrace and support hazardous sites, sewage sludge and weeds.[21] Even though food hovers as a tenuous possibility rather than a present reality in this research, *Revival Field* models complex futures for both agriculture and mining. The land pinned in the sights of the field's target design may someday return to the agricultural fold. The metals seeping into the plants may someday be collected and repurposed. Meanwhile, as with *Notations*, a contained ecosystem produces fields of colour and chemical changes while performing resilience and survival.

Newton Harrison looks back on the *Survival Pieces* and marvels that 'life is tough'.[22] *Revival Field* evokes a similar sense of wonder at the tenacity of living things. At once a map, a mine and an agricultural potential, the site brings digestion to the fore despite the inedibility of its 'crops'. Instead of producing consumable crops, this field consumes itself. Hyperaccumulator plants absorb metals, cultivating a green gut. Resilient plants labour to ingest, reclaim and revive the land.

Writing about 'plant-thinking', Michael Marder challenges humans to 'respect the obscurity of vegetal existence', and to resist instrumentalising plants, for 'whenever human beings encounter plants, two or more worlds (and temporalities) intersect: to accept this axiom is already to let plants maintain their otherness, respecting the uniqueness of their existence'.[23] How does *Revival Field* encounter plants, and does the project respect 'the obscurity of vegetal existence'? At first it would appear that *Revival Field* does precisely what Marder would ask us to resist: plants are instrumentalised in capacities that far exceed typical and familiar uses of plants as edible crops. Here, plants mine metals from the earth, to be extracted by a burning process that would obliterate their vegetal bodies and worlds. But even as this apparent instrumentalisation blazes through the hyperaccumulators populating *Revival Field*, complications emerge. Cadmium and zinc destabilise the taxonomic divisions between plants and people. Advocating pollution means attending to what plants and people share as well as to that which differentiates us. In this case, heavy metals bind the vegetal world to the human world, with phytoremediation processes allowing humans to encounter plants on their own terms. These plants do what humans cannot. Witnessing the processes of hyperaccumulation and the benefits produced within *Revival Field* soils allows the plantworld to encroach on the human world. Humans are forced to adopt the plant's point of view, the plant's timeframe and the plant's startlingly different and strange capacities. Chin knows the processes set in place in *Revival Field* will extend beyond his human life. He has begun to understand the long duration of plant temporality alongside the value of vegetal obscurity, and the radical difference of plants.

Revival Field's plant participants extract zinc and cadmium, with iron-rich sludges binding if not extracting lead. Of course zinc, cadmium and lead are just three of many metals polluting soil and entire ecosystems. Natalie Jeremijenko's *Cross(x)Species Adventure Club* designs and serves meals that facilitate encounters between animals, plants, humans and pollutants across diverse materials, ecosystems and places. Menus include oil spill cocktails, meals for geese and people and investigations of water, soil and *terroir*. Jeremijenko's *Lures* work with mercury by way of slippery edibles that she declares to be 'the iconic example' of the project.[24] The *Lures* tackle food and heavy metals at once, addressing the 'bio-amplification that happens through the food web'.[25] As with *Revival Field*, pollutants promote cross-species encounter. Designed at the aesthetic intersection of bait worms, gelatinous gin and tonics, gummy candies and luminescent fishing lures, the long, segmented squiggles are one of few foods designed to be ingested by both humans and fish: 'Small actions to feed the fish can aggregate significant effect, augment the nutritional resources and improving the health of fish, aquatic ecosystems and humans.'[26] While many humans eat fish, rarely do humans think about eating with fish, or sharing their food with fish. The *Lures* seat people and fish at the same table. By eating together, fish and humans make an accumulative positive change in mercury contamination across species.

Fish absorb mercury from water contaminated by human activities like burning coal, but they excrete methylmercury much more slowly than they take it in. Metals then lodge in human tissues after eating methylmercury-contaminated fish.[27] *Lures* clear mercury from human and fish bodies with chitinase, a chelating agent that transforms heavy metals into harmless salts.[28] This approach recalls Rufus Chaney's extra intervention in more recent *Revival Fields* with *in situ* remediation using iron-rich sludges. Both the sludge and the chitinase bind pollutants. While hyperaccumulator plants absorb metals, sludge and chitinase transform and contain them, removing pollutants from bioavailability. The entire food web benefits from binding metals in this way: the *Lures* are edible across species, and if fish can eliminate heavy metals more quickly, humans are exposed to one fewer source of potential methylmercury contamination.

Like Chin and the Harrisons, Jeremijenko embraces collaboration in her projects. In her *Environmental Health Clinic* research platform, of which *Cross(x)Species Adventure Club* is a public-facing component, her collaborations are both scientific and culinary. In addition to drawing on her own expertise with engineering, biological and physical sciences, Jeremijenko developed a research partnership with Professor Mark Bain's laboratory at Cornell while designing the *Amphibious Architecture* component of this larger research initiative. Culinary collaborations are equally important. Jeremijenko works with molecular gastronomist Mihir Desai to engineer foods with unusual and sophisticated textures and tastes. Jeremijenko explains that molecular gastronomy matters to the project because 'the complexity of the tastes that people are experiencing makes them open to hearing about these complex systems and processes that they would otherwise not be motivated to hear'.[29] The subtleties enabled by molecular gastronomy model a corporeal, gastric approach to technologies of environmental remediation.

For example, Jeremijenko and Desai employ 'reverse spherification' to craft 'terrestrial bubbles' from beetroot. These red blobs are structurally akin to egg yolks: the egg's vitelline envelope separates the yolk from the white, and when the teeth pierce this thin, transparent membrane, the yolk's liquid centre spills out. The beetroot bubbles are held together by a chemical version of the vitelline membrane. Reverse spherification is relatively simple to do at home, and has consequently become one of the most familiar molecular gastronomy techniques.[30] The process requires no special equipment, just two chemicals, which can be purchased in packaged commercial kits. The reaction between calcic gluconolactato and sodium alginate precipitates a thin membrane (like the egg's vitelline envelope) around small quantities of liquid (in this case beetroot). Bubbles on land prompt diners to question 'terrestrial versus aquatic environments and where we fit in'.[31] Encapsulating the subterranean beetroot in a buoyant bubble allows these foods to enact the complexities Jeremijenko alludes to above: the fragile technology of the thin chemical membrane destabilises fundamental assumptions about earth, water and air. A simple soup would not have the same effect. The techniques of molecular gastronomy allow for greater complexity and more nuanced outcomes as *Cross(x)Species Adventure Club* makes practical and theoretical interventions into ecological systems.

Elsewhere I argue that in the *Cross(x)Species Adventure Club*, 'digestion becomes a way of figuring the landscape and encountering animals, plants, and environmental systems, with molecular gastronomy techniques providing metaphoric and literal mechanisms for imagining how bodies and landscapes interrelate'.[32] Like the Harrisons's salty brine shrimp and Chin and Chaney's hyperaccumulators, hungry for heavy metals, the chelating agents in Jeremijenko's *Lures* bind methylmercury by way of digestive processes. Our bodies are bodies of water; what transpires inside our skin compels consideration of other bodies of water large and small: ponds, puddles, rivers and oceans. Dining alongside fish makes us aware of what we share. Contaminants like mercury entangle organisms across species, enmeshing us in the messy knot that Donna Haraway, after Isabel Stengers, calls 'multispecies cosmopolitics': 'relations in multispecies cosmopolitics work by indigestion and infection, rather than reproduction'.[33] While interspecies reproduction is a challenge, infection flourishes between biologically different bodies (both physically, as with human and feline immunodeficiency virus, and figuratively, as with mercury-infected humans and fish). We might further extend the idea that methylmercury contamination acts as a form of infection, with the *Lures* in turn acting as an inoculation, hijacking digestion to spread metal binding capacities: remediation as virus. For Stengers, cosmopolitics as a modality hinges on obligation:

> The term 'cosmopolitics' introduces what is neither an activity, nor a negotiation, nor a practice, but the mode in which the problematic copresence of practices may be actualized [...] It is a form of asymmetrical reciprocal capture that guarantees nothing, authorizes nothing, and cannot be stabilized by any constraint, but through which the two poles of the exchange undergo a transformation[34]

When a *Cross(x)Species Adventure Club* participant eats a lure or feeds one to a fish, human and fish experience the copresence and obligation Stengers describes. *Cross(x)Species Adventure Club* promotes generative indigestion, transforming pollutants into salts and salts into waste. *Lures* embrace bioremediation technologies while activating multispecies interaction by building on existing complexities.

Like many of the *Cross(x)Species Adventure Club* projects, the *Lures* are amphibious, working across land, water and waterways to write the latest chapter in a long history of the effects of industrialisation on watersheds. This history begins with Ellen Swallow Richards and the Normal Chlorine Map. In 1903, after spending several decades sampling water all over the United States, Richards concluded that '[i]t is hard to find any place in the world where the water does not show the effect of human agencies'.[35] Early ecologists like Richards understood that there was no pure, uncontaminated wilderness. In order for sanitation to become a framework for environmental remediation, there would have to be a widespread understanding of something akin to our contemporary conception of the anthropocene: the 'effects of human agencies' are pervasive, but incremental changes are possible. Richards had advocated for water sanitation for many years before embarking on the largest survey of the United States municipal water system, 'The Great Sanitary Survey', in 1887. This two-year study was conducted in partnership with the aptly named Dr Thomas Drown. Richards found that she was 'still the number two man', even though she was solely responsible for more than 100,000 analyses of water samples taken from nearly every body of water in Massachusetts.[36] The most visible outcome of the study was the Normal Chlorine Map, which plots the chlorine levels of over 40,000 water samples. The amount of chlorine in the water and the divergence of a sample from 'normal' together indicate the amount of man-made pollution contaminating the water. The Great Sanitary Survey and the Normal Chlorine Map led to the first water quality legislation in the United States with the inclusion of minimum standards for public drinking water in the Massachusetts State Sanitary Code.[37]

Although it might be overstating the case to argue that the Normal Chlorine Map created the environmental movement in the United States, projects like it facilitated early air and water quality legislation, and set a standard for documenting and visualising industrial pollution. Maps and surveys reveal a mutable landscape of human agencies and toxins, suggesting that intervention is possible. Artists may not have been central to the Great Sanitary Survey of 1887, but in the decades following these early efforts to track environmental pollution, survey data have become increasingly important for artists, and mapping and surveying have emerged as critical practices for contemporary artists and scientists alike.[38] Among artists and art collectives dedicated to mapping practices that visualise everything from surveillance to environmental pollution, the Center for Genomic Gastronomy makes a particularly apt intervention in the context of this chapter's focus on bioremediation. The Center's recent *Smog Tasting* project combines surveys, sample testing, and taste testing to give edible form to the 'human agencies' Richards was plotting at the turn of the twentieth century. Although this project does not actively ameliorate

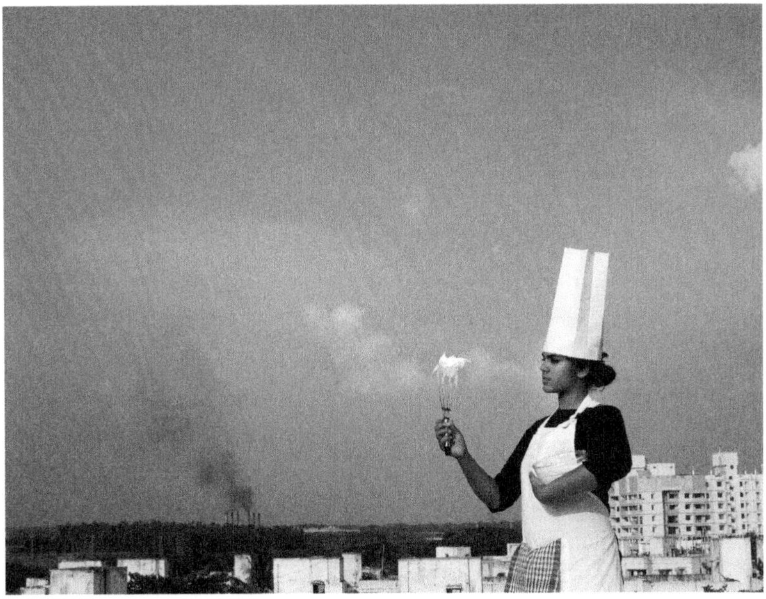

Figure 10.1 The Center for Genomic Gastronomy, *Smog Tasting* (2010). Photo courtesy of the Center for Genomic Gastronomy.

air pollution, the Center's focus on sampling, taste and mapping produces the same kinds of environmental identity shifts provoked by projects like Jeremijenko's and Chin's.

Zack Denfeld and Cathrine Kramer founded the Center for Genomic Gastronomy in 2010, and together with Emma Conley and Heather Julius they have produced recipes and events that consider local ecologies, unusual flavours and new food technologies. Their menus often resonate with *Cross(x)Species Adventure Club* menus. The *Loci Food Lab* series, for example, presents 'bioregional food research in public' a recent event in Oregon used survey responses to customise a snack from nine local ingredients, including marionberry fruit leather and 'Western Red Cedar Simple Syrup'.[39] The Center also considers species extinction with projects like *De-Extinction Deli* and *Colony Collapse Cuisine* and imagines overlaps between aquarium and culinary technologies with recipes for *Glowing Sushi*. Their 2010 project *Smog Tasting* maps air pollution with meringues, capturing the atmospheres of some of the world's most polluted cities using egg foams, which are 90 per cent air. Each city has a dedicated spice mix, with ingredients combined in ratios that reflect the particles found in that city's air. When observing air pollution, vision and smell tend to be prioritised – we look at a hazy cityscape or inhale fumes on the street. By shifting our perception of air pollution to the taste buds, *Smog Tasting* meringues reorganise human senses. The meringues are samples, data that generate a map of global air quality using taste as the key index.

All of these projects extend from sanitation's nuanced understanding of ecology's place in the home and its earliest interventions into environmental policy. Taken together, the projects in this chapter share the Center for Genomic Gastronomy's investment in using modelling, mapping and visualisation to approach bioremediation by way of unexpected sense-encounters. With *Notations*, we see salinity before we taste it. With *Revival Field*, we encounter phytotime, and are obliged to adapt to plant rhythms that endure after our human hearts stop beating. With *Cross(x) Species Adventure Club*, we share our food with fish, experiencing intimate connection between bodies of water bounded by skin, scales, riverbanks and continents. Food and eating situate human bodies in nested relationships that knot together the home and the landscape. Our taste buds allow us to access very small particles enfolded in meringues as well as very large problems and paradigm shifts that 'reverberate at least on continental levels'.[40] Together, these works combine food and ecology to promote an expanded posthuman sensorium.

11

Plumpiñon

> Are we fully aware that we are about to connect, and thus transform through multiple and partly unpredictable acts of reciprocal capture, histories that, on Earth and until now, were distinct?
>
> Isabelle Stengers[1]

Plumpiñon

2 tablespoon piñon nuts, ground into paste
1 teaspoon vitamin powder
1 teaspoon powdered sugar
1 teaspoon granulated sugar

Mix together to form a paste. For additional comfort, fill a small plastic bag, snip a corner, and suck the paste from the bag.[2]

~

What do we eat when we eat technology? My recipe for *Plumpiñon* puts this book's theories of hybrid studio-kitchen production, multispecies entanglement and interdisciplinary approaches to food and eating into practice. Developed with precise outcomes in mind, therapeutic foods are edible technologies. As with Subject P's oatmeal experiments with survival food, the eating body becomes a data point, a result. Manipulating the technological underpinnings of ready-to-eat therapeutic foods becomes a process of teasing apart postcolonial politics, family histories and expanding notions of the local.

Plumpy'nut ('plump' + 'peanut') is a peanut-based humanitarian aid food developed by André Briend and licensed by Nutriset, a French company specialising in therapeutic food. Plumpy'nut requires no water or refrigeration, making it ideal for extreme environments that provide little support for malnourished people. Used most often with infants and children in Africa (especially in Niger and Malawi), Plumpy'nut has also played a role in tsunami relief efforts. Writing for *the New York Times*, Michael Wines witnessed Plumpy'nut distribution and consumption in Maradi, Niger. He is quick to point out that Plumpy'nut's humble ingredients should not be underestimated:

> Plumpy'nut, which comes in a silvery foil package the size of two grasping baby-size hands, is 500 calories of fortified peanut butter, a beige paste about as thick as mashed potatoes and stuffed with milk, vitamins and minerals. But that is akin to calling a 1945 Mouton Rothschild fortified grape juice.[3]

Wines' fine wine analogy suggests that Plumpy'nut allows simple ingredients like peanuts and powdered milk to transcend their everyday function and participate in something grander. By making Plumpy'nut into wine, he shifts its purpose from fulfilling desperate need to enjoying luxurious pleasure.

Perhaps *New York Times* readers do need creative metaphors in order to understand how Plumpy'nut works and why it matters, especially in Niger. Situated in the Sahara desert, landlocked Niger suffers severe droughts, leading to desertification and famine. Plumpy'nut is a rich, quick intervention for malnourished children, replacing local starvation foods like 'bitter berries', the fruit of the Anza bush (*Boscia senegalensis*). Outside observers often understand eating bitter berries as 'pica', or the pathologised practice of consuming non-food: media coverage of past and present famine in Niger emphasises that Anza berries are 'inedible' when famine does not compel their consumption.[4] Refeeding, pica and starvation foods escape conventional understanding of eating and nourishment. Such practices reveal bodies at their limits, whether those bodies are enmeshed in subsistence living or the exploration of alien landscapes. The peanut's familiarity to Western palates allows for an easier understanding of Plumpy'nut's value. We see this familiarity at work with backpacker favourite Justin's Peanut Butter, which serves the 'highest quality natural and organic ingredients' in small foil packets strongly resembling those used by Nutriset.[5] Even if the foil packet delivery system is unusual, nut butters offer more accessible flavours than the counterintuitive and consequently undesirable bitterness of the Anza bush's berries.

Plumpy'nut hails from France, a former colonial power in Niger and many other African nations. Nutriset's entanglements with colonial pasts haunt its present status as a copyright-holding entity that exerts cultural and economic pressures on local producers. Although Plumpy'nut production happens locally, producers are beholden to Nutriset for the patented vitamin mixture that is added to the peanut butter base. Sending engineered laboratory foods to Niger ignores the potential of local starvation foods like bitter berries.[6] The well-travelled peanut acted as starvation food for the pre-Columbian South American civilisations responsible for its domestication.[7] Under what conditions might the Anza berry and the acorn thrive and travel? Given the historical confluence of postcolonial politics and identity, multispecies engagement with hardy flora like Anza and Piñon, and my own family's story and geography, *Plumpiñon* comes to represent both a formal inquiry into the structure and sociality of humanitarian aid food and an investigation of reciprocal relationships between settler colonies and indigenous peoples.

As a remembrance of local famine foods around the world, the on-going performance series and research initiative I call *Plumpiñon* promotes one such famine food native to the US Southwest. The native piñon tree's seeds function much like Plumpy'nut and the Anza bush. New Mexico and the Southwest US present a richer edible landscape than Niger, but this region also suffers lean times and famines. *Pinus edulis*, the Colorado pine, has historically been – and continues to be – an important source of nutrition, fuel and cultural connection in both Colorado and New Mexico, where it is the state tree. The nutritious and storable seed keeps for long periods of time, and the strong, durable wood of the pine tree is valued for construction and smells sweet when burned. Piñon pine forests can be found in Colorado, southern Wyoming, eastern and central Utah, northern Arizona, New Mexico and the Guadalupe Mountains in westernmost Texas.[8] Also known as Colorado pinyon or two-needle pinyon, the trees frequently hybridise with other pine tree varieties and rely on the pinyon jay (*Gymnorhinus cyanocephalus*) to disperse their seeds.[9] Piñon forests are old and can appear sparse. This drought-tolerant tree grows slowly and is not very tall, reaching around 30 feet at most.[10] Slow growth means that pine nut crops are increasingly limited.

When I bought pine nuts in California, I was most likely buying nuts sourced from China. Among the varieties of nut on the shelf, pine nuts are often the most expensive. Nuts sourced from the US Southwest could cost up to twice as much as imports. While visiting my family in Colorado, I might be given a small plastic bag containing a handful of piñon nuts gathered from a friend's property, but I would not be able to purchase locally-grown piñon in the store. Even when passed from hand to hand, piñon seeds circulate in plastic of one form or another. Recalling the plastic skins that stretch over industrial and local seeds alike, I package *Plumpiñon* in sealed plastic bags made from recycled materials, and my recipe contains pine nuts, sugar, protein powder and, optionally, powdered milk. *Plumpiñon* is different from Plumpy'nut in both its taste and texture even though its plastic packet closely follows the delivery system developed for Plumpy'nut.

I first experimented with squishy bags of paste in January of 2009, when I offered a public demonstration of *Plumpiñon* production. While talking about the cultural and family histories that connect me with the piñon nut, I prepared *Plumpiñon* for about a dozen people and served it in small plastic bags obtained from a bead shop. The presentation prompted me to pursue alternative ways of delivering *Plumpiñon*. Heather Paxson's portmanteau '*microbiopolitics* [...] call[s] attention to the fact that dissent over how to live with microorganisms reflects disagreement about how humans ought to live with one another'.[11] The recipe development and packaging process for *Plumpiñon* became microbiopolitical interventions. The paste and its delivery systems question the logic of Plumpy'nut, its corporate origin points and the ethics of copyrighting food.

Sterility and concern about contaminating microorganisms are at the heart of Plumpy'nut packaging choices. My homemade equivalents are

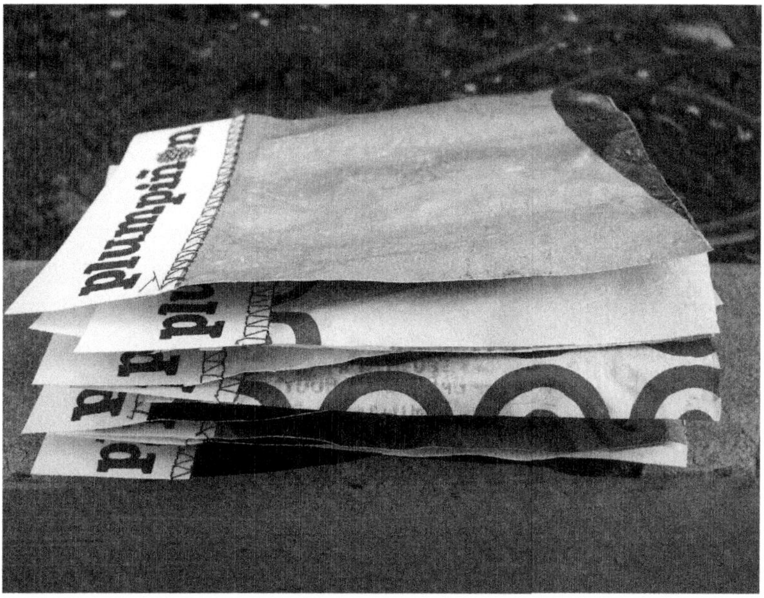

Figure 11.1 Lindsay Kelley, *Plumpiñon* (2010).

'sealed' in such a way as to invite rather than guard against, in Paxson's words, 'human encounters with the vital organismic agencies of bacteria, viruses, and fungi'.[12] *Plumpiñon*'s imperfect containers are made from a material called 'fused plastic'. This production method resists Plumpy'nut's mode of mass-produced packaging: fused plastic is easily made with an iron, parchment paper and polyethylene shopping bags, a material that has been the subject of several ecological design interventions in recent years.[13] To give the motley collection of fused plastic bags a sense of visual unity, I developed a logo for *Plumpiñon* based on the Plumpy'nut logo. Plumpy'nut uses a peanut as an apostrophe, signalling the nut butter content, and giving peanuts a kinetic, active presence in the product. My logo replaces the 'o' in *Plumpiñon* with a pine cone, and uses a similarly friendly typeface. Handmade bags, small-scale production and extremely expensive ingredients combine to critically, even subversively, reflect on Plumpy'nut's method of branding and Nutriset's corporate ownership.

Airdrop packaging resembles plastic tarps and fused plastic, with its durable, rip-stop nylon with reinforcement threads. In both process and product, the activity and visual culture of air dropping makes a number of commodity chains and intercultural knots visible. Airlifts reach inaccessible places to deliver, in the military context, everything from tanks to supplies to propaganda leaflets. Humanitarian aid airdrops from UN planes are designed to distinguish materials and planes from military counterparts. Airdrops pass judgment about the safety and accessibility of the terrain where supplies are dropped. Looking down from the plane, removed from the ground where the parachutes will land, aid stays in the air, cocooned in materials and markings specially developed to withstand the physical and cultural distance created by flight.

Transforming a staple starvation food from the Southwest United States into a relief food used around the world comments on both poverty and luxury. While many eaters are familiar with the pine nut as an ingredient in pesto or salad, people living in the Southwest remember surviving on the pine nut as recently as 75 years ago. Birdie Jaworski interviewed one such person in her essay 'Pride in piñon'.[14] Jane Yazzie is a Navajo woman in her 80s living in Bernal, New Mexico, where she harvests some 50 pounds of piñon nuts each autumn, roasting them by hand before making traditional foods like biscochitos (the New Mexico state cookie). According to Vernon Mayes and Barbara Bayless Lacy's *Nanise': A Navajo Herbal*,[15] the piñon pine is involved in nearly every Navajo ceremony, either as seed, wood or pitch. Also used as medicine, food, fuel and in trade networks, the piñon is a vital element of Navajo foodways and culture.[16] Harvest rights for piñon seeds entangle land, people and labour: the Bureau of Land Management in Nevada 'has closed some traditional gathering areas to commercial harvest to ensure that treaty obligations' with Native Americans are honoured,[17] while in Colorado, Arizona, and New Mexico, 'Navajos are the major commercial pickers of pinyon nuts', their skill far surpassing other commercial pickers.[18]

Harvesting piñon nuts is arduous work. First the tree must be shaken, then the nuts and cones and needles that fall from the tree must be sorted and stored. Yazzie remembers the piñon nut helping her family survive the dust bowl:

> I remember eating piñon for days. For weeks. It was all we had. We kept these big sacks filled with the last piñon harvest. My

> mother used to tell me that we were like piñon. We sometimes have a hard shell around us when times are difficult, but our insides are always sweet.[19]

As both a starvation food and a metaphor for the body, the piñon's pale sweet meat mirrors the flesh of those consuming it. Yazzie remembers how her grandmother stored piñon nuts in clay pots and in holes in the ground. Isolating foods in clay, earth, plastic or foil helps nuts and seeds survive conditions that make other food spoil. Yazzie's town, Bernal, sits near a mountain called Starvation Peak. The legend behind the naming of the mountain involves a detachment of colonial Spanish soldiers. Having been chased to the top of Starvation Peak, the soldiers chose to die there rather than be captured. The Southwest is full of stories and places that mark colonial communication breakdowns, failed reciprocity, unsuccessful negotiations and the violence of starvation.

My family, mostly living in Denver, Colorado, shares entanglements with *Pinus edulis*. My great-grandfather grew up in the small town of Madrid, New Mexico. When he was a child, Madrid was a coal mining company town, where Mexican-Irish alliances were common, and complicated. Today, Madrid bills itself as 'a recovered ghost town that now exists as an art destination',[20] with a population of 204 in the 2010 census, only seven of whom identify as 'Hispanic or Latino'.[21] For my family, whiteness is a mutable, contested identity. My great-grandfather contributed to our cultural blanching by changing his name from Narvaez to Kelley, taking his mother's name.[22] Taking one's mother's name is both a traditional Spanish/Latino practice and a way to negotiate how a name might affect opportunities. Of course, my great-grandfather would not have identified as a man 'of colour'. He would have positioned himself somewhere inside a messy knot that included multiple national, ethnic and racial identities, all of which were moving targets of the US government as land and people in this region were introduced to still-changing and ill-fitting taxonomies.

Although my family does not share Yazzie's memories of surviving on piñon nuts, and we do not gather the seeds, even recreationally, our family history and myths are connected to the seeds nonetheless. When my great-grandfather died, one of his cousins visited my grandparents and told us that my great-grandfather had a living brother and that remaining estranged brother kept the name Narvaez. This name, my distant cousin explained, was inherited from Pánfilo de Narváez, a Spanish conquistador who wrecked his ships in the Gulf of Mexico in 1528, launching the expedition that would make Cabeza de Vaca the first European (and his slave Esteban the first African) to travel though the American Southwest. My great-grandfather's cousin believed us to be cultural kin to this conquistador (despite his immediate shipwreck and death upon arrival), and therefore Spanish. Many New Mexican families have a similar story. If I accept our version of this story, then my family is intimately connected with the piñon and its peoples' first experience of Spanish presence in the Southwest. These seeds have thus come to be associated with colonial activity as much as with native survival tactics. In *The Piñon Pine: A Natural And Cultural History* – an interdisciplinary approach to the cultural life of the piñon tree that includes recipes and picking instructions – forest biologist Ronald Lanner writes that if the seeds did help de Vaca survive, 'the piñon pine casts a long shadow indeed, for when the tattered survivors reached Mexico, their arrival touched off a series of historically important expeditions into New Mexico'.[23]

Scholars have debated Cabeza de Vaca's exact route for centuries, and the pine nut has helped provide evidence of de Vaca's path through the Southwest. De Vaca created a multispecies community on his journey, becoming enmeshed with the seeds to such a degree that his detailed descriptions of nuts with paper-thin shells have allowed biogeographers to determine his route based on the prevalence of the piñon variety *Pinus remota*, or 'paper-shell Piñon'.[24] Perhaps de Vaca's intimacy with piñon was an 'act of reciprocal capture' or a precocious moment of 'cosmopolitics'.[25] Learning how to eat piñon coincides with a turn in de Vaca's journey: he is able to perceive and narrate the 'copresence of practices' as never before.[26] de Vaca describes his experiences with piñon nuts in the context of exchange:

> They gave us many beads and many hides of cows, and they loaded all those who came with us with some of everything they had. They ate prickly pears and pine nuts, and there are throughout that land small pines, and the cones of them are like small eggs, but the pine nuts are better than those of Castile,

> because they have a very thin hull. And when they are green they grind them and make them into lumps and eat them in this manner, and if they are dry they grind them with the hulls and they eat them as a powder.[27]

Without local people, perhaps de Vaca would have recognised the pine nut's edibility (he refers to similar nuts in Spain) but he certainly would not have understood their versatility or significance to the Southwestern diet and landscape: the people who teach him how to eat this landscape facilitate de Vaca's multispecies meal.

Writing about Mungo Park's *Life and Travels of Mungo Park in Central Africa*, Mary Louise Pratt defines 'reciprocity' as 'the desire to achieve equilibrium through exchange'.[28] James Clifford reminds us that '"reciprocity" is a translation term, whose meanings will depend on specific contact situations'.[29] Here I attempt such a translation, reading reciprocity for Park against the tension at play in de Vaca's *Journey*. Torn between base survival and 'the greatest non-reciprocal non-exchange of all time: the Civilizing Mission', de Vaca, like Park, becomes 'that creature in whose viability and authenticity his readers may have longed to believe: the naked, essential, inherently powerful white man'.[30] De Vaca's description of the pine nuts prefaces his emerging role as an unlikely faith healer, a career he embarks on as a way of deepening the reciprocity he earlier sought to engage through trade. For example, de Vaca performs surgery on a man with an arrow point embedded in his chest, giving de Vaca and his companions 'a very great reputation'.[31] De Vaca deals in flesh and life: as piñon sustains him, he sustains the human flesh of those who gave him piñon flesh to eat. He works with touch, faith and occasional tangible surgeries. Trading life for life, pulling life from faith, from thin air, from nothing at all, shows that de Vaca recognised the life-giving capacity of piñon pines.

Unlike de Vaca, a group of later explorers, the Donner party, would not be captured by piñon and would die as 'pure nomads' after flirting with survival cannibalism. For Stengers, 'the only one who is dangerous, irremediably destructive or tolerant, is someone who believes himself to be "purely nomadic", because he can only define his practice in contrast to all the others and, regardless of his good intentions, can only define others in terms of tolerance'.[32] Cannibalism is not the inevitable end result of a purely nomadic strategy, but the particular cannibalism enacted by the Donner party in the midst of nurturing vegetation, would be an excellent example of pure nomadism. For some reason, even though Indians had probably demonstrated how to gather and eat the nuts, the Donner party turned to cannibalism before eating piñon. Ronald Lanner wonders why the Donner party chose human flesh over nutmeat:

> Why didn't the hard-pressed emigrants collect piñon nuts along their route through Nevada? [...] They could have gathered a stock of piñon nuts that would have seen them through the winter, but they did not try. Their journals never mention piñons. Were they ignorant of the value of pine nuts, despite the experiences of earlier travelers? Did their hostility to the Indians they encountered along their route discourage the Indians from offering piñons in trade?[33]

De Vaca introduces pine nuts in the context of his evolving efforts to achieve a state of reciprocity with the human and plant cultures he encounters on his journey. That human flesh could be seen as food while nut flesh could not reveals the cultural specificity of pine nuts and their limited legibility as food. Lanner's suspicion that systemic and specific hostilities may have blocked communication with the Donner party provides a telling counterpoint to de Vaca's experience, and illuminates the distinction between capture and captivity. Hostility, fear and intercultural violence promote captivity (as with the Spanish detachment starving at the top of Starvation Peak), while hunger without tolerance makes space for multispecies productivity.

Accepting gifts of food and clothing precipitates the exchange that turned the tide of de Vaca's journey, allowing him to move among Indians freely and productively and to be physically and gastronomically captured by a multispecies landscape. The Donner party never experienced this turn or engaged such concepts of reciprocity, instead keeping to themselves, unable to perceive the food sources all around them. The limits of Nigerien foodways resemble the limits of Yazzie's sacks of piñon seeds in that scarcity has forced, with varying degrees of comfort, 'reciprocal capture' across species boundaries. The Nigerien may be captured by *Boscia senegalensis* in much the same way that Yazzie and de Vaca were captured by *Pinus remota*.

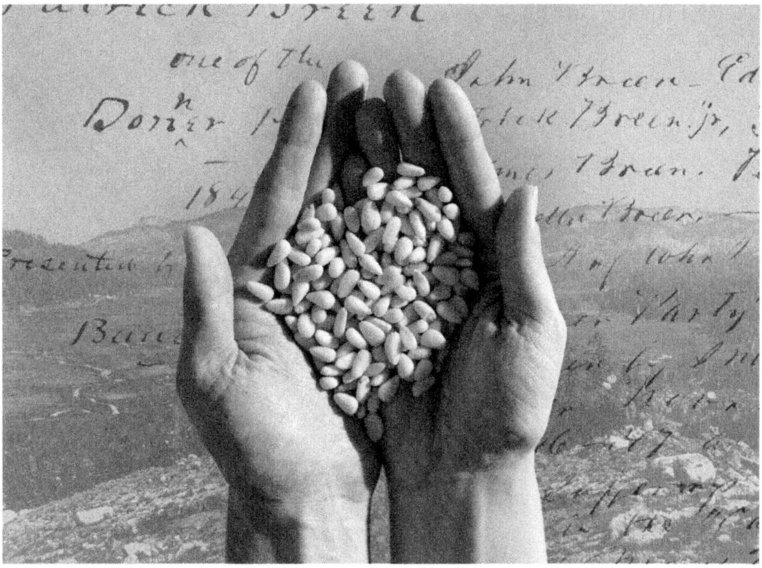

Figure 11.2 Lindsay Kelley, *Starvation Seeds* (2009).

Nut flesh stands in for the flesh of people, both living and cannibalised, and also traces a history of reciprocal relationships between Indians and whites, colonial, contemporary, and newly minted, like my family. Acceptance or rejection of pine nuts has meant acceptance of certain types of cultural life, as well as acceptance of survival and life itself. The Donner party's dietary speciesism demonstrates a non-reciprocal relationship with the land: captivity rather than capture. If starvation can be negotiated with, human animals must walk the desert hoping to be taken in by the land, thus figuring capture as a multispecies entanglement that nurtures and sustains.

Epilogue: Dysphagiac

Mixed Vegetables or Broccoli

3 Servings
4 cups of frozen mixed vegetables (corn, green beans, peas, lima beans, etc.). This is about ½ of a large bag.
3 cups of broth, juice or cooking water
Salt and Pepper to taste
Optional: 2 tablespoons of olive oil

Cook vegetables for time indicated on package. Drain cooking water into a bowl and put cooked vegetables in the processor. Pulse 10 times. Process for one minute continuously. Add 3 cups of cooking liquid or other broth. Add salt/pepper to taste and oil if desired.[1]

~

Returning to the question 'What is food?' I conclude with a case study of the edges or fringes of food. Medical necessity has forced many caretakers into creative amateur engagements with food science and dietetics. This epilogue documents *Dysphagiac*, a project that responds to a family medical crisis with the same critical attention to technology, domesticity and living systems that animates this book project and my practice in general. This work put many of my theories about food, art, and resistance into visceral practice – my awareness of bioart's entwined history with Home Economics and public amateurism facilitated both the speculative design process I relate in the following paragraphs and a more nuanced understanding of my family and our enmeshment in the American health care industrial complex.

For people living with dysphagia, the inability to swallow, defining food stretches both the imagination and the membrane separating starvation from digestion. The dysphagiac[2] turns inward, finding dining companions in plasticities of the stomach, potentials of the gastric feeding tube and the interior rumblings of the humanimal microbiome. If the dysphagiac has convenient access to health care, she might have a gastrointestinal tract tube which pumps liquid nutrition directly into the stomach. Without such technological intervention, physical exercises that encourage the swallowing reflex are the only hope for avoiding starvation. There are many conditions that may produce dysphagia, including stroke, cerebral palsy, Parkinson's disease, cancer and injury to the oesophagus. We enter the world on a liquid diet of baby formula, often manufactured by Nestlé, and we sometimes leave the world on the same liquid diet. From the summer of 2008 until his death in September 2009, my grandfather ate his meals by way of a GI tube. He never stopped attending to food as food, and politely (dangerously) accepted food on offer from people who did not understand his condition. If he did manage to swallow anything, it slipped into his lungs, causing complications like pneumonia.

The inability to swallow marks a turn away from humanist conceptions of sustenance. People in my grandfather's position learn to ignore food and eating by mouth. My grandfather found his 'nil by mouth' status to be both easy and difficult. Not eating was easy because he was never hungry; four times a day the stomach tube provided him with enough nutrition to avoid hunger pangs. Not eating was difficult because he missed the oral pleasures of food and longed for the experience of eating, which pleased both his senses and his sociality. My grandmother kept food out of their apartment kitchen and ate all her meals in the dining room of their independent living community. She continued to do this for years after his death, until her own death in 2014. Her cupboards emptied alongside her husband's oesophagus, and were never restocked.

What do tubes eat? My grandfather dedicated their spare bathroom to feeding the tube and tracking his weight. Cans filled the closet. Stacked by the case, the cans evoke a bomb shelter's supply room, ready to survive the apocalypse. My grandfather's cans were manufactured by Nestlé, and were flavoured, although their contents only touched his tongue as burps or vomit. This is an empty larder, a simulation of sustenance. If these cans keep at room temperature for years, how can they substitute for fruits and vegetables that would rot in a matter of days? These cans imagine the stomach as space of bare life and survival, only able to ingest the minimum, most sterile, most preserved. Defender of food Michael Pollan advises us to 'shop the peripheries of the supermarket'.[3] Pollan's imagination of the stomach compares the organ to a compost heap, to earthly rot, cycles of decay and fecundity. Wishing to return my grandfather's stomach to him in all of its messy potentiality, I researched other ways to feed the tube. I looked for ingredients that did not focus on survival and preservation, ingredients absent from the inert centre aisles of the supermarket.

The first and best resource I found was a website published by Lucy Duperon called 'Lucy's Real Food'. Duperon's fiancé was unable to swallow after surgery for cancer, and she developed the recipes to satiate his hungry tube. Her recipes are acts of resistance, critiquing the fallout shelter fare pushed on them by big pharma. Duperon's efforts were rewarding. She writes:

> Almost immediately after switching to real food, my fiancé began regaining weight (while on the feeding tube!) and his digestion and elimination became much closer to normal […] Our doctors have been first surprised and then impressed […] Using real food in the tube creates a sense of normalcy during mealtimes, can include favorite foods and, importantly, allows us to correct for nutritional deficiencies such as anemia and other depletions.[4]

Enteral feeding always results in weight loss (my grandfather lost over 20 pounds in less than a year). Duperon's success with homemade liquid nutrition made me hopeful that I could help my grandfather gain some weight and enjoy feeding the tube. In winter 2009, I prepared Duperon's recipes for 'Mixed Vegetables and Tamales'. I froze ten servings of the purèed food and carried it from San Francisco to Denver in my luggage. Below-freezing temperatures in Denver ensured that the liquid food was frozen solid on arrival.

This was my first act of feeding as resistance. My process was successful: tamales reached their liquid state effortlessly, according to instructions, and I found the recipes easy to follow. The food I prepared nine months before he died was sitting in my grandparents' nearly empty freezer when I left Denver, and gone when I returned three months later. My experiment failed. My grandfather did not ingest the food I made and continued to consume the canned Nestlé food until he died of complications related to Parkinson's disease (a primary complication being the tube itself). My grandfather's death manifested almost entirely in his throat and lungs. He coughed, he could not breathe, he could not swallow; he choked. It seemed that eating or drinking would soothe him. But he couldn't eat or drink.

My grandmother was a formidable cook. She supervised three-day-long chili roasting endurance tests, enchilada preparations that occupied the entire house and back porch, Thanksgiving time trials that involved waking throughout the night to change oven temperatures. Her mother put food on the table for 15 children every single day for an incomprehensible number of years. Her mother gave birth to all of those children, and more who died. My great-grandmother's body must have been in a constant state of near collapse while her kitchen was under constant threat from dough-pinchers inside and the Great Depression outside (hobos, dogs, rats, secret codes of hospitality and neglect carved into fences). My great-grandmother was cooking through, baking through, working through. From the excessive reproduction and generation of her joyful and exhausting childhood, my grandmother learned to cook with anything, to buy the best ingredients she could afford and to prefer oil to butter. She learned to save, to scrape every last crumb of sifted flour from her waxed paper surface, then use that paper five more times before discarding it. She learned to stock her pantry just so. She knew which corner of which cupboard housed which spice, which pan was on which shelf, which cereal box was only a quarter full.

The instructions for consuming Lucy's liquid food recipes are simple. Heat gently to body temperature. Stir. Feed the tube. But my grandmother would not feed the tube. She declared that my grandfather 'wasn't allowed to eat that'. That someone so confident in the kitchen could not take up the tube as another appliance to be leveraged for love speaks to how powerful and even threatening prescriptions and prescriptive discourse can be. I continue to make and promote liquid food, three years after my grandfather's death. My grandfather never benefited from this process. I hope that someone else might. Neither my grandfather nor his nurses formally objected to the idea; instead, my grandmother's will and pain dictated his course of treatment. Her arthritis prevented her from working with batter and dough (the stand mixer we bought her doesn't do a good enough job). A fracture in her cervical spine precipitated her move to a graduated care facility. Given her failing culinary dexterity, I understand why she refused to learn how to assist him with the feeding tube. Grandpa was not an eater anymore: he was a patient. What if my grandfather's doctor had been one of Duperon's, one of 'our doctors'? What might have been different? Could anyone have helped my grandmother feed the tube?

I have prepared liquid food for many people, none of whom used a tube to consume it. During public tastings, I found that liquid cherry pie was the most popular item on offer. We appreciate sweet liquids: smoothies, milkshakes, soft drinks, sugared tea and coffee. I share Duperon's recipes through tasting events because they teach us how to move our functioning bodies into illness and death, performing complex conceptual work by turning the body inside out. More critically, I prepare and converse about these foods because of the importance of real nutrition for people living with GI tubes. Doctors prescribe canned formula, not real food, so dysphagiacs tend to eat canned food. People who feed tubes and stomachs at once rebel.

Tubes feed a range of would-be eaters. Stomach and throat tubes 'refeed' people who are capable of swallowing (anorectics, bulimics, hunger strikers). In this case, when one could but will not eat through the mouth, enteral feeding becomes a constitutive outside made material as a technology of the body. Feeding the tube challenges our conception of what it means to eat, what food should resemble and contain and whether and how chewing and swallowing contribute to meals. Tubes (usually through the nose and throat, not the stomach) also provide a means for women, often brides, to undergo medically supervised crash diets.[5] At some point the refed anorectic and the crash diet bride meet one another on a continuum of disordered eating and complex networks of self-perception. They meet in the tube.

Liquid food evokes visions of the future, often a dystopian future, where even the young and beautiful are supported by intravenous or intraintestinal food sources. Speculative fiction and film familiarise us with the tube. These futures prepare us for illness and death, for the time when we cannot eat, when we cannot swallow, when we must take in and feed the tube, the catheter, the intravenous drip. In the film *The Matrix* (1999), humans are overgrown foetuses, suspended in pink embryonic liquid and supported by a system of umbilical tubes locked into the nervous Apollonian spine, not the navel. In the animated film *WALL-E* (2008), humans evolve into obese blobs with stumpy appendages, their fatty tissues kept plump by suckling endless liquid milkshakes through plastic straws. In Margaret Atwood's novel *Oryx and Crake*, precocious bioengineer Crake gives his bioengineered Children digestive systems that resemble those of rabbits, requiring a diet of leaves and fresh greens, which are then recycled as soft caecotrophs, 'semi-digested herbage, discharged through the anus and re-swallowed two or three times a week'.[6]

In *The Sleeper Awakes*, H.G. Wells predicts a startlingly happy liquid food 'with a pleasing faint aroma and taste and a quality of immediate support and stimulus'.[7] The body has lost the ability to chew but has gained the olfactory pleasure of supportive, neutral nutrients. The question of pleasure in bodily breakdown preoccupied the late Bob Flanagan. Artist, masochist and exiled poster boy for the Cystic-Fibrosis Foundation, Flanagan feeds the tube by way of a re-engineered 'Visible Man' toy. The hacked Visible Man ejaculates, vomits and dribbles imitation excrement made from wood putty, with the whole system powered by leftover IV pumps. *Visible Man* (1992) plays with the same tensions between inside and outside as the preparation of liquid food. Wells's mild beverage joins with the slow dribbles of the *Visible Man* to reveal how the processes hidden by our skin might sloppily and messily emerge from the body.

Although focused on what comes out rather than what goes in, Flanagan's *Visible Man* nonetheless feeds the tube. Flanagan demonstrates the medicalisation of the body by turning its insides out. The oesophagus becomes a plastic tube. The anus becomes a plastic tube. The urethra becomes a plastic tube. The anatomical paths of liquids and the body's own capacity to produce liquids run parallel to Flanagan's process of finding inorganic substitutes for organic substances. Kirby Dick's documentary about Flanagan, *Sick: The Life and Death of Bob Flanagan, Super-Masochist* (1997), contains a great deal of footage detailing the process behind *Visible Man*. Infused with humour and frustration, Flanagan's difficulties with glue-semen recipes stand in for his difficulties with his own body. The medical doll surrogate protests an efficient, tidy, future world where the palate is irrelevant, infantilised, or fundamentally alien to our current modes of consumption.

Inspired by liquid food and tube futures, *Dysphagiac* contains a sculpture, a research initiative and a portable workshop centred on a countertop appliance designed with laminar flow cabinetry principles. Enfolding a blender in a red latex drape that evokes the exploration of endoscopy footage, the stomach appliance draws on tissue culture laboratory protocols to conceptualise the kitchen and the stomach as spaces of sterility, infection, fecundity and medical intervention. Taking the gastric feeding tube as a fertile site for investigating ecologies of gut bacteria and corporeal morphism, *Dysphagiac* shows how multispecies symbiosis and domestic technoscience co-produce digestive tracts. The appliance goes beyond previous tasting formats to question the place of the dysphagiac in the home. Excluded by the kitchen and dining room, dysphagiacs become patients and eating becomes a medical procedure. Creating counter space for a liquid food appliance returns the dysphagiac to the kitchen.

Referencing safe rooms and bodily cavities, *Dysphagiac* houses demonstrations, including handwashing techniques, recipe development and preparation and tasting events that compare liquefied perishable foods with the prescription fare typically ingested by tube-fed patients. In a 2013 tasting at the University of Toronto, chef Joshna Maharaj created a multi-course meal that we cooked and liquefied in a large industrial kitchen that supplies University dining halls. Maharaj works with local hospitals to improve liquid nutrition programmes, providing a hopeful example of how health care systems might embrace alternative modes of ingestion. Along with a group of researchers affiliated with the Jackman Humanities Institute's 'Food' postdoctoral and residency programme, I prepared and tasted a range of liquid foods, including spaghetti, salad, berry cobbler and a vanilla flavoured prescription formula. A later incarnation of this

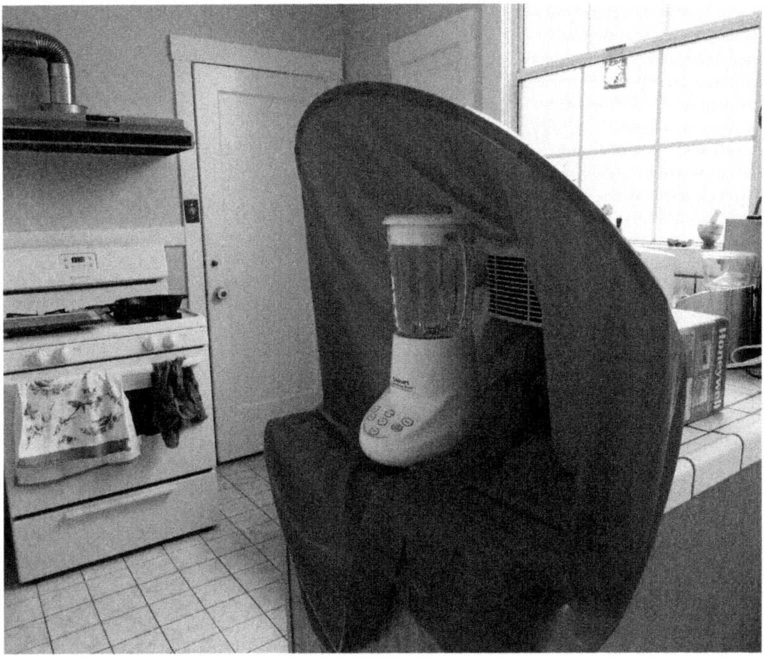

Figure 12.1 Lindsay Kelley, *Dysphagiac* (2013).

workshop at the exhibition 'Feral Experimental', hosted by the University of New South Wales Department of Art and Design, focused on vegetables and banana bread.

Whether in my kitchen at home or in a speculative future, preparing liquid food, fresh or industrial, turns the body inside out and renders subcutaneous technologies of the body concrete and visible outside the membrane of the skin. I found that *Dysphagiac* both grounded and contextualised the alternate histories for bioart that I examine throughout this text. Liquid food preparation encapsulates many of the relations that animate the arguments in this book – the project joins bioart works dwelling in the porous boundaries between humans and animals, organisms and machines, scientific and amateur modes of engagement. These works, like the Home Economics project, are fundamentally about care. How do we feed ourselves and those we love? How do we critically confront the injustices of everything from industrial agriculture to domestic labour in the home? Methodologies developed in response to these questions combine the material with the imaginary. In the case of *Dysphagiac*, a speculative kitchen appliance undoes industrial agricultural and pharmaceutical systems by reimagining the physiology of eating. By preparing stomach contents outside the body, the blender and strainer perform roles normally filled by chewing, swallowing and passing food through the oesophagus. Preparing liquid food forces one to imagine the mechanical work performed by the body, and all the ways that body is entangled in knots of profit, medicine and normalising discourses of health. Our food lands in the stomach rather than the lungs with only a thin flap of flesh promising the correct trajectory. Each swallow astonishes me.

Notes

Introduction: What is Food?

1 Betty Crocker, *Betty Crocker's Picture Cookbook* (Minneapolis, MN: Wiley and General Mills, 1950), p. 125.

2 Jacques Derrida with Jean-Luc Nancy, '"Eating well", or the calculations of the subject: an interview with Jacques Derrida', in Eduardo Cadava, Peter Connor and Jean-Luc Nancy (eds), *Who Comes after the Subject?* (New York: Routledge, 1991), p. 115. In her essay 'Parting bites', Donna Haraway likens Derrida's 'excessive responsibility' and confusion over the 'who' of 'who comes after the subject?' to a 'giving up' of human exceptionalism. Donna Haraway, *When Species Meet* (Minneapolis, MN: Minnesota University Press, 2008), pp. 285–301.

3 Derrida and Nancy, '"Eating well"', p. 115.

4 Expansively defined by Jans Hauser as a 'proliferating mutant term', 'bioart', an enmeshment of 'biological' or 'biology' and 'art', has travelled extensively in contemporary art discourse. Bioart finds materials and process inspiration in the laboratory, incorporating wet biology into sculptural, time-based, sometimes edible work. See Jens Hauser, 'Taxonomy of an etymological monster', in Gerfried Stocker and Christine Schopf (eds), *Ars Electronica: Hybrid – Living in Paradox* (Ostfildern-Ruit: Hatje Cantz, 2005), pp. 182–93.

5 Beatriz da Costa and Kavita Philip's edited collection *Tactical Biopolitics: Art, Activism, and Technoscience* (Cambridge: MIT Press, 2008) includes writing from artists discussed in this book (Critical Art Ensemble, Oron Catts, Ionat Zurr, Claire Pentecost, da Costa herself) as well as from other makers and thinkers whose work contributes to bioart conversations (subRosa, Jens Hauser, Kathy High, Paul Vanouse). The book brings biologists, artists and theorists into dialogue about both tactical media's current manifestations and legacy and the relevance of Foucault's concept of 'biopower' to recent '(un)disciplinary' engagements with media and technology' (ibid., p. xviii). Artist Eduardo Kac edited *Signs of Life: Bioart and Beyond* (Cambridge: MIT Press, 2007), which also brings together artists and philosophers but proposes a different set of questions. Kac is concerned with the creatures of bioart – 'hybrids, clones, mutants, synthetics, and transgenics' (see 'Introduction', ibid., pp. 1–12) – as well as situating bioart within art history, a project da Costa and Philip do not pursue, given

that tactical media has always resisted and thrived outside of art history. Also of interest: Donna Haraway's writing about Patricia Piccinini in *When Species Meet*, Cary Wolfe's writing about Alba and Sue Coe in *What is Posthumanism?* (Minneapolis: Minnesota University Press, 2009) and Joanna Zylinska's discussion of bioart in *Bioethics in the Age of New Media* (Cambridge: MIT Press, 2009). Steve Baker's *Artist | Animal* (Minneapolis: Minnesota University Press, 2013). See Allison Carruth's "Culturing Food: Bioart and In Vitro Meat" for a different perspective on many of the same texts I address in this book. Carruth works through food studies and literature to contribute to a shared commitment to broadening understandings of bioart histories and presents. Allison Carruth, 'Culturing Food: Bioart and In Vitro Meat', *parallax* 19/1 (2013), pp. 88–100. Sheryl Vint's paper at SLSA 2009, 'Disembodied Cuisine: Eating Well in The Mad and Oryx and Crake', was another moment of happy synchronicity. Available at http://www.litsciarts.org/slsa09/program.php (accessed 31 August 2015).

6 Michael Pollan, *In Defense of Food: An Eater's Manifesto* (New York: Penguin, 2008) and *The Omnivore's Dilemma: A Natural History of Four Meals* (New York: Penguin, 2006); Morgan Spurlock, *Super Size Me* (Kathbur Pictures, 2004); Aaron Woolf, *King Corn* (ITVS, 2007); Warren Belasco, *Meals to Come: A History of the Future of Food* (Berkeley, CA: California University Press, 2006); Hervé This, *Molecular Gastronomy: Exploring the Science of Flavor* (New York: Columbia University Press, 2005); Susan Marks, *Finding Betty Crocker: The Secret Life of America's First Lady of Food* (New York: Simon & Schuster, 2005); Arlene Voski Avakian and Barbara Haber (eds), *From Betty Crocker to Feminist Food Studies: Critical Perspectives on Women and Food* (Boston: Massachussetts University Press, 2005); Mary Drake McFeely, *Can She Bake a Cherry Pie? American Women and the Kitchen in the Twentieth Century* (Amherst, MA: Massachusetts University Press, 2000); Sidney Mintz, *Tasting Food, Tasting Freedom: Excursions into Eating, Culture, and the Past* (Boston, MA: Beacon, 1996).

7 With Lynn Turner, I co-edited *parallax* 66: 'bon appétit' (19/1 (2013)). A special issue of *Environmental Communication: A Journal of Nature and Culture* 4/3 (2010), edited by Andy Opel, Joseé Johnston and Richard Wilk, focuses on 'Food, culture and the environment: communicating about what we eat'. Eva Hayward, Helena Pedersen and other researchers in the HumAnimal group at the Centre for Gender Research at Uppsala University organised a conference called 'Meet animal meat'. All of the food at the conference was vegan, prompting individual consideration of meat and dairy products by setting new norms. Available at http://www.genna.gender.uu.se/conferences-events/conferences-workshops/meet-animal-meat/ (accessed 7 February 2014). 'Zoontotechnics (animality/technicity)' at the Centre for Critical and Cultural Theory, Cardiff University, addressed 'philosophical-ethical re-valuations of the "animal" and renewed reflections on various aspects of technology

and technics, both within and beyond the emerging framework of posthumanism'. Available at http://www.cardiff.ac.uk/encap/newsandevents/events/conferences/zoontotechnics.html (accessed 7 February 2014).

8 Jane Bennett, 'The agency of assemblages', *Vibrant Matter: A Political Ecology of Things* (Durham, NC: Duke University Press, 2010), pp. 20–38.

9 A number of studies have described spaces similar to art labs. Caroline Jones and Peter Galison's book *Picturing Science Producing Art* (New York: Routledge, 1998) traces intersections between art studios, laboratories and experimental venues from the Enlightenment onward. Within the book, see especially David Freedberg, 'Iconography between the History of Art and the History of Science: art, science, and the case of the urban bee', pp. 272–96; Svetlana Alpers, 'The studio, the laboratory, and the vexations of art', pp. 401–17; and Bruno Latour, 'How to be iconophilic in art, science, and religion?', pp. 418–40.

10 The book *Beyond Productivity: Information Technology, Innovation, and Creativity* explores computer science through an interdisciplinary lens, asking how computer science benefits from interactions with other communities, art and design being of primary importance. *Beyond Productivity* investigates a space called the 'studio-laboratory' that resembles the art lab: 'studio-laboratories [...] bring together different domains of knowledge, research, and practice–thus combining the artist's (or designer's) studio with the scientist's (or engineer's) laboratory'. William J. Mitchell, Alan S. Inouye, and Marjory S. Blumenthal (eds), *Beyond Productivity: Information Technology, Innovation, and Creativity* (Washington DC: National Academies Press, 2003), p. 118.

11 Projects like Hans Ulrich Obrist and Barbara Vanderlinden's *Laboratorium* exhibition and accompanying catalogue open both art spaces and science laboratories to the public, offering public displays and performances alongside visiting hours at laboratories around Antwerp. Hans Ulrich Obrist and Barbara Vanderlinden (eds), *Laboratorium* (Antwerp: Dumont, 1999).

12 See one of the oldest recorded cookbooks, dating from 1390, *The Forme of Cury*. Available at http://www.pbm.com/~lindahl/foc/ (accessed 7 February 2014).

13 See Suzanne Anker, *The Molecular Gaze: Art in the Genetic Age* (Cold Spring Harbor, NY: Cold Spring Harbor Laboratory Press, 2004), Gerfried Stocker and Christine Schöpf's contributions to *Hybrid: Living in Paradox* (Ostfildern: Hatje Cantz, 2005), Ingeborg Reichle, *Art in the Age of Technoscience: Genetic Engineering, Robotics, and Artificial Life in Contemporary Art* (New York: Springer, 2009), Pramod Nayar, 'The new monstrous: digital bodies, genomic arts and aesthetics', *Nebula* 4/2 (2007), pp. 1–19 and Marvin Heiferman, Carole Kismaric and Ian Berry, *Paradise Now: Picturing The Genetic Revolution* (Saratoga Springs, NY: Tang Teaching Museum, 2001). Controversially, the *Paradise Now* exhibition was revealed to have been sponsored by biotech corporations, prompting a critical response from participating artists, led by Natalie Jeremijenko.

14 Ionat Zurr, 'Growing semi-living art', PhD thesis, University of Western Australia 2009, p. 145. Available at http://theses.library.uwa.edu.au/adt-WU2009.0125/ (accessed 7 February 2014).

1 Subject P: Embodying Home Economics

1 H.C. Sherman, J.C. Winters, and V. Phillips, 'Efficiency of oat protein in adult human nutrition', *The Journal of Biological Chemistry* 39/1 (1919), pp. 53–62, p. 57.
2 Robert Mitchell, *Bioart and the Vitality of Media* (Seattle, WA: Washington University Press, 2010), p. 26.
3 'Timeline of the New York State College of Home Economics, 1900–1969.' Available at http://rmc.library.cornell.edu/homeEc/timeline.html (accessed 9 March 2014). For another timeline of Home Economics history, see Virginia B. Vincenti, 'Chronology of events and movements which have defined and shaped Home Economics', in Sarah Stage and Virginia B. Vincenti (eds), *Rethinking Home Economics: Women and the History of a Profession* (Ithaca, NY: Cornell University Press, 1997), pp. 321–30.
4 Ellen Richards, *The Chemistry of Cooking and Cleaning* (Boston, MA: Estes & Lauriat, 1882).
5 For more on the Rumford Kitchen, see 'The Rumford Kitchen exhibit at World's Columbian Exposition, Chicago, 1893: excerpts from *Report of the Massachusetts Board of World's Fair Managers*'. Available at http://libraries.mit.edu/archives/exhibits/esr/esr-rumford.html (accessed 9 March 2014).
6 For MIT's account of Richards' career, see 'Ellen Swallow Richards and MIT'. Available at http://libraries.mit.edu/archives/exhibits/esr/esr-mit.html (accessed 9 March 2014). For more information about the Women's Laboratory at MIT, see Elizabeth Durant, 'A lab of their own', *MIT Technology Review*. Available at http://libraries.mit.edu/archives/exhibits/esr/esr-womenslab.html (accessed 9 March 2014). Also see Sarah Stage, 'Ellen Richards and the social significance of the Home Economics movement', in Stage and Vincenti, *Rethinking Home Economics*, pp. 17–33.
7 Stage, 'Ellen Richards', p. 21.
8 Mary Drake McFeely, *Can She Bake a Cherry Pie? American Women and the Kitchen in the Twentieth Century* (Amherst, MA: Massachusetts University Press, 2000), pp. 34–5.
9 For a study of teachers and students practicing Home Economics in the UK, see Dena Attar, *Wasting Girls' Time: The History and Politics of Home Economics* (London: Virago, 1990). For more about how Home Economics interfaced with the home and parenting, see Julia Grant, 'Modernizing mothers: Home Economics and the parent education movement, 1920–1945', in Stage and Vincenti,

Rethinking Home Economics, pp. 55–76. Also see a pie chart of Home Economics college graduates' careers, 'Home Economics graduates'. Available at http://rmc.library.cornell.edu/homeEc/3careers/chart.html (accessed 9 March 2014).

10 'What were practice apartments?' Available at http://rmc.library.cornell.edu/homeEc/cases/apartments.html (accessed 9 March 2014).

11 For more on recipes, objectivity, and science, see Janet Theophano's analysis of William Eamons in Janet Theophano, *Eat My Words: Reading Women's Lives through the Cookbooks They Wrote* (New York: Palgrave, 2002), pp. 89–90, 290, and William Eamons, *Science and the Secrets of Nature: Books of Secrets in Medieval and Early Modern Culture* (Princeton, NJ: Princeton UP, 1994).

12 Nancy Berlage, 'The establishment of an applied social science: Home Economists, Science, and Reform at Cornell University, 1870–1930', in Helen Silverberg (ed.), *Gender and American Social Science: The Formative Years* (Princeton, New Jersey: Princeton University Press, 1998), p. 190.

13 Megan Elias elaborates on the amateur/professional divide along gendered lines: 'Although home economists attempted to align themselves with the experts, the culture more generally classified anything to do with the kitchen and with women as amateur.' Megan Elias, *Stir it Up: Home Economics in American Culture* (Philadelphia, PA: Pennsylvania University Press, 2010), p. 27. For a detailed history of how women interfaced with scientific institutions in the United States, see Margaret Rossiter, *Women Scientists in America: Struggles and Strategies to 1940* (Baltimore, MD: Johns Hopkins University Press, 1982) and *Women Scientists in America: Before Affirmative Action, 1940–1972* (Baltimore: Johns Hopkins University Press, 1995).

14 For a more extended reflection on how 'female experts' came to 'dictate 'standards of home living' to women,' see Stage, 'Ellen Richards', p. 32. Also see Ruth Schwartz Cowan, *More Work for Mother: The Ironies of Household Technology from the Open Hearth to the Microwave* (New York: Basic Books, 1983) and Barbara Ehrenreich and Deirdre English, *For Her Own Good: Two Centuries of the Experts' Advice to Women* (New York: Anchor, 1978).

15 Cowan, *More Work for Mother*, pp. 201.

16 Christine Frederick, *Selling Mrs Consumer* (New York: The Business Bourse, 1929).

17 Carolyn Goldstein, 'Part of the package: Home Economists in consumer products industries, 1920–1940', in Stage and Vincenti, *Rethinking Home Economics*, pp. 271–96, p. 273.

18 See Jeanne Paris, *Your Future as a Home Economist* (New York: Richards Rosen, 1964) and Velma Phillips, *Home Economics Careers for You* (New York: Harper & Brothers, 1962).

19 Phillips, *Home Economics Careers*, p. 196.

20 Ibid., p. 225.

21 Paris, *Your Future as a Home Economist*, p. 81.

22 These titles were selected from 469 thesis titles collected in the United States Department of Agriculture and Office of Education, *Titles of Completed Theses in Home Economics and Related Fields in Colleges and Universities of the United States, 1954–1955* (Washington DC: US Government Printing Office, 1955).
23 Phillips, *Home Economics Careers*, p. 224.
24 Sherman, Winters and Phillips, 'Efficiency of oat protein', pp. 53–62.
25 Paris, *Your Future as a Home Economist*, pp. 101–2.
26 See 'The kitchen laboratory', in McFeely, *Can She Bake a Cherry Pie?*, pp. 34–50.
27 One result of this new direction in Home Economics may be seen in the curriculum for the Department of Consumer and Family Studies/Dietetics in the College of Health and Human Services at San Francisco State University. Coursework emphasises child development, food science and preparation for teaching. 'Consumer and Family Studies/Dietetics, San Francisco State University'. Available at http://cfsd.sfsu.edu/ (accessed 10 March 2014).
28 Quoted in Stage and Vincenti, *Rethinking Home Economics*, p. 1.
29 Attar, *Wasting Girls' Time*, p. 2.
30 For a discussion of Beecher's manuals and their significance to popular understanding of domestic economies, see Susan Strasser, 'Redeeming woman's profession', in her *Never Done: A History of American Housework* (New York: Holt, 1982), pp. 180–201.

2 Chicken Heart Soup

1 'Food for thought (and for your face)'. Available at http://food4thoughtand4urface.wordpress.com/tag/oryx-and-crake/ (accessed 29 January 2014).
2 Phil Ross, *Biotechnique* (San Francisco, CA: Yerba Buena Center for the Arts, 2007).
3 Hannah Landecker, *Culturing Life: How Cells Became Technologies* (Cambridge MA: Harvard University Press, 2007), p. 15.
4 For example, the Dada movement roughly parallels the careers of both Carrel and Fell and the advancement of tissue culture as a methodology. Of course Dada's politics are markedly different from Carrel's, who infamously promoted eugenics. If differing political commitments can be overlooked, Dada captures the era's attention to process over product, and a commitment to time, duration and event that echoes Bergson's theoretical interventions.
5 Ionat Zurr, 'Growing semi-living art', PhD thesis, University of Western Australia 2009, pp. 62–3. Available at http://theses.library.uwa.edu.au/adt-WU2009.0125/ (accessed 7 February 2014).
6 Susan Squier, 'Life and death at Strangeways: the tissue culture point of view', in Paul Brodwin (ed.), *Biotechnology and Culture: Bodies, Anxieties, Ethics* (Bloomington, IN: Indiana University Press, 2000), pp. 27–52, p. 31.

7 Carrel quoted in Landecker, *Culturing Life*, p. xi.
8 For a meditation on time, its varied aspects and the importance of physiological time for living organisms and 'the growth energy of tissues', see Alexis Carrel, 'Physiological time', *Science* 74/1929 (1931), pp. 618–21.
9 Landecker, *Culturing Life*, p. 105.
10 Hannah Landecker locates the chicken heart's final hour in Leonard Hayflick's discovery that 'normal somatic cells in culture consistently divide for a set number of generations [...] chicken heart cell cultures go through thirty-five doublings at the most – far shorter than the time the immortal chicken heart "lived". Thus it seemed impossible that Carrel's culture could have been composed of normal chicken cells. Hayflick concluded that the chick embryo extract preparation Carrel used as a nutrient medium provided new viable embryonic cells at each feeding.' Landecker, *Culturing Life*, pp. 166–7.
11 Henri Bergson, *Matter and Memory* (New York: Zone, 1991), p. 210.
12 Oron Catts and Ionat Zurr, 'Semi living art', in Eduardo Kac (ed.), *Signs of Life* (Cambridge, MA: MIT Press, 2007), p. 245.
13 Carrel, 'Physiological time', p. 618.
14 Robert Mitchell, *Bioart and the Vitality of Media* (Seattle, WA: Washington University Press, 2010), pp. 31, 67.
15 A number of popular articles tracked Carrel's unusual aesthetic sensibilities, notably 'Men in black', *Time*, 13 June, 1938, pp. 40–3. Andrés Horacio Reggiani collects critical perspectives on Carrel under the heading 'Laboratory work as theater'. Andrés Horacio Reggiani, *God's Eugenicist: Alexis Carrel and the Sociobiology of Decline* (New York: Berghahn, 2007), pp. 40–3.
16 Ted Howard and Jeremy Rifkin, *Who Should Play God? The Artificial Creation of Life and what it Means for the Future of the Human Race* (New York: Delacorte, 1977), pp. 106.
17 Alexis Carrel and Charles Lindbergh, *The Culture of Organs* (New York: Hoeber, 1938), p. 81.
18 Ibid.
19 Hannah Landecker, 'Immortality, *in vitro*: a history of the HeLa cell line', in Brodwin, *Biotechnology and Culture*, pp. 53–74, p. 56.
20 Zurr, 'Growing semi-living art', p. 62.
21 Landecker, *Culturing Life*, p. 33.
22 David Friedman, *The Immortalists: Charles Lindbergh, Dr Alexis Carrel, and their Daring Quest to Live Forever* (New York: HarperCollins, 2007), p. 63.
23 This and information about Huxley can be found in James Gunn (ed.), *The Road to Science Fiction*, vol. 2: *From Wells to Heinlein* (Lanham, MD: Scarecrow, 2002), pp. 144–5.
24 Julian Huxley, 'The tissue culture king', ibid., p. 147.
25 Ibid., p. 154.

26 Ibid.

27 Emphasising the religious, ritualised aspects of tissue culture production recalls an instance of eating overlapping with subsidiary lives: consuming the extended body of Christ in Catholic mass. The extended body is a rich metaphor for understanding transubstantiation. Huxley's story, with its ritual atmosphere and ancestor worship, conjures this association vividly, transposing transubstantiation on the king and his tribe and making Hascombe's project akin to that of missionaries. Huxley was often critical of the Church, being in favour of birth control and even championing the work of French Jesuit-palaeontologist Pierre Teilhard de Chardin after the Jesuits prohibited the publication of his work. Making Hascombe an agent of Catholicism would reinforce the idea that he is a decadent madman. See Julian Huxley, *Memories* (New York: Harper & Row, 1970).

28 Huxley, 'The tissue culture king', p. 166.

29 Hannah Landecker examines the category of race in discourse about the HeLa cell line, describing a tendency toward 'personification of the cell line' (Landecker, *Culturing Life*, p. 142). Race enters the HeLa story nearly 20 years after the cells were harvested, and racial origin stories continued to dominate the conversation until recently, when the biotech industry changed the terms of the debate: 'endless reproduction and worldwide distribution remain part of the story of Lacks's immortality, but the metaphors have become those of the growth of capital while those of miscegenation and contamination have retreated into the background' (ibid., p. 173). See also Landecker, 'Immortality, *in vitro*'.

30 Huxley, 'The tissue culture king', p. 156.

31 Monica Bakke, 'Zoe-philic desires: wet media art and beyond', *parallax* 14/3 (2008), pp. 21–34.

32 Margaret Atwood, *Oryx and Crake* (New York: Doubleday, 2003), p. 202.

33 Ibid.

34 Winston Churchill, 'Fifty years hence', *Popular Mechanics*, March 1932, pp. 390–7.

35 Atwood, *Oryx and Crake*, pp. 242, 315.

36 Ibid., p. 242.

37 Ibid., p. 205.

38 Ibid., p. 158.

39 Sidney Mintz, *Tasting Food, Tasting Freedom: Excursions into Eating, Culture, and the Past* (Boston, MA: Beacon, 1996) p. 7.

40 Donna Haraway, *When Species Meet* (Minneapolis, MN: Minnesota University Press, 2008), p. 268.

41 Ibid., p. 267. For video footage of chickens unable to walk due to genetic manipulations of muscle mass, see the video Compassion Over Killing, '45 days in hell: the life and death of a broiler chicken'. Available at http://www.chickenindustry.com/cfi/documentary/ (accessed 10 March 2014).

42 Harold McGee, *On Food and Cooking: The Science and Lore of the Kitchen* (New York: Scribner, 2004).

43 The 2000 version of the Avigenics website has been removed from the Internet Archive's Wayback Machine, but the text quoted here was taken from that archive. Today, Avigenics former website redirects to the webpage of Synageva BioPharma. 'Synageva BioPharma'. Available at http://www.avigenics.com/ (accessed 3 February 2015). For more about how chicken eggs play a role in developing vaccines, see Landecker, *Culturing Life*, pp. 110–21.

44 Fell delivered a series of lectures about the tissue culture point of view between 1936–8. Squier, 'Life and death at Strangeways', pp. 31–3.

3 Domestic Computing

1 This recipe appears in a number of Helen Corbitt and Neiman Marcus cookbooks. I found this exact version with the headline 'The Ultimate Dallas Salad', *D Magazine* (September 2013), p. 55.

2 A US War Department pamphlet with the commanding title *You're Going to Employ Women* recommends that hiring managers 'interpret machinery operation in terms of household and kitchen appliances'. The text goes on to explain that women workers are 'pliant, dexterous, finger-nimble, good at repetitive tasks'. *You're Going to Employ Women* (Washington DC: US War Department, 1943).

3 Jennifer Light, 'When computers were women', *Technology and Culture* 40/3 (1999), pp. 455–83. People are no longer referred to as computers; Light explains that 'a "computer" was a human being until approximately 1945. After that date the term referred to a machine, and the former human computers became "operators"'. Ibid., p. 469.

4 Joanna Zylinska, *Bioethics in the Age of New Media* (Cambridge, MA: MIT, 2009), p. 176.

5 Ellen Lupton, *Mechanical Brides: Women and Machines from Home to Office* (New York: Cooper Hewitt, 1993), p. 9.

6 James Tomayko, 'Electronic Computer for Home Operation (ECHO): the first home computer', *IEEE Annals of the History of Computing* 16/3 (1994), pp. 59–61.

7 Judy Wajcman, *Feminism Confronts Technology* (University Park, PA: Pennsylvania State University Press, 1991), pp. 89, 95.

8 Tomayko, 'Electronic Computer for Home Operation (ECHO)', p. 61.

9 Glenn Infield, 'A computer in the *basement*?', *Popular Mechanics* (April 1968), pp. 77–9, 209, 229.

10 Ibid., p. 78 (emphasis in original).

11 For a poetic analysis of how women's bodies, advertising and machine utopias collide, see Marshall McLuhan, *The Mechanical Bride: Folklore of Industrial Man* (Corte Madera, CA: Ginko Press, 2002).

12 For a nuanced examination of household technology and labour, see Ruth Schwartz Cowan, *More Work for Mother: The Ironies of Household Technology from the Open Hearth to the Microwave* (New York: Basic Books, 1983).

13 Infield, 'A computer in the *basement*?', p. 78 (emphasis in original).

14 The Computer History Museum, 'Bytes for bites: the kitchen computer', n.d. Available at http://www.computerhistory.org/revolution/minicomputers/11/362 (accessed 23 January 2014).

15 Ibid.

16 Dag Spicer, 'If you can't stand the coding, stay out of the kitchen: three chapters in the history of home automation', *Dr Dobbs' Portal*, 12 August 2000. Available at http://www.ddj.com/184404040 (accessed 23 January 2014).

17 Cowan, *More Work for Mother*, p. 18.

18 Valerie Aurora, 'Valerie Aurora meets the Honeywell H316 Kitchen Computer', n.d. Available at http://valerieaurora.org/kitchen.html (accessed 4 February 2015).

19 Wajcman, *Feminism Confronts Technology*, pp. 106, 107.

20 Margaret Morse, *Virtualities: Television, Media Art, and Cyberculture* (Bloomington, IN: Indiana University Press, 1998), p. 128 (emphasis in original).

21 Genevieve Bell, Mark Blythe and Phoebe Sengers, 'Making by making strange: defamiliarization and the design of domestic technologies', *ACM Transactions on Computer-Human Interaction* 12/2 (2005), pp. 149–73.

22 'Nintendo Wii controlled smarthome', *Liquidice's Nintendo Wii Hacks*, 23 December 2006. Available at http://wiihacks.blogspot.co.uk/2006/12/nintendo-wii-controlled-smarthome.html (accessed 23 January 2014).

23 Ian Fried and Melanie Austria Farmer, '3Com lets Audrey out the door', *CNET*, 17 October 2000. Available at http://news.cnet.com/3Com-lets-Audrey-out-the-door/2100-1040_3-247152.html (accessed 23 January 2014).

24 Whirlpool, 'Smart appliances featuring smart home technology', n.d. Available at http://www.whirlpool.com/smart-appliances/ (accessed 23 January 2014).

25 Megan Wollerton, 'LG HomeChat lets you text with your appliances', *CNET*, 30 December 2003. Available at http://ces.cnet.com/8301-35306_1-57616358/lg-homechat-lets-you-text-with-your-appliances/ (accessed 23 January 2014).

26 Kevin Ashton coined the term 'internet of things' to promote the idea that machines need ways of accessing information that go beyond human input. If more objects were equipped with RFID sensors, for example, a networked refrigerator could alert a household when food has expired and place an order for more. In the context of the spam-sending kitchen appliances, 'internet of things' evokes a shadow world of networked devices operating without sufficient oversight. Kevin Ashton, 'That "internet of things" thing', *RFiD Journal*, 22 June 2009. Available at http://www.rfidjournal.com/articles/view?4986 (accessed 23 January 2014). One of many news articles about spam appliances: 'Fridge

sends spam emails as attack hits smart gadgets', *BBC News*, 17 January 2014. Available at http://www.bbc.co.uk/news/technology-25780908 (accessed 23 January 2014).

27 Anna Munster, *An Aesthesia of Networks* (Cambridge, MA: MIT Press, 2013), p. 184.

4 *Semiotics of the Kitchen*: Feminist Food Art

1 Martha Rosler, *Semiotics of the Kitchen* [1975] (Cambridge, MA: MIT Press, 2004).

2 Linda Montano, 'Food & art', *High Performance* 4 (Winter 1981–2), pp. 45–54, p. 45.

3 Judy Chicago, *The Dinner Party: A Symbol of Our Heritage* (Garden City, NY: Anchor Books, 1979), p. 35.

4 Nina Felshin (ed.), *But is it Art? The Spirit of Art as Activism* (Seattle, WA: Bay Press, 1995), p. 10.

5 Although Home Economics seemed socially conservative, it helped to generate Cornell's contemporary women's studies programme with its feminist perspective on gender. In January 1969, before the name change, Cornell hosted a controversial Intersession Program on Women that was sponsored by the College of Home Economics. The conference sparked an Ithaca chapter of the National Organization of Women (NOW) and the first collegiate 'female studies' course offered by the Department of Human Development and Family Studies. 'Why the change to human ecology?' (2001). Available at http://rmc.library.cornell.edu/homeEc/cases/namechange.html (accessed 14 April 2012).

6 Arlene Voski Avakian and Barbara Haber (eds), *From Betty Crocker to Feminist Food Studies: Critical Perspectives on Women and Food* (Boston, MA: Massachusetts University Press, 2005), p. vii.

7 Mira Schor, *A Decade of Negative Thinking: Essays on Art, Politics, and Daily Life* (Durham, NC: Duke University Press, 2009), p. 232.

8 Schor, *A Decade of Negative Thinking*, p. 232.

9 Pope's broadside reads 'Add 2 cups of Warhol and a stick of Nauman and serve!' Ibid., p. 239.

10 For a large collection of FLUXUS recipes/scripts/scores, see Ken Friedman, Owen Smith and Lauren Sawchyn (eds), *Fluxus Performance Workbook*, digital supplement to *Performance Research* 7/3 (September 2002). Available at http://www.thing.net/~grist/ld/fluxusworkbook.pdf (accessed 18 March 2014).

11 Montano, 'Food & art', p. 50.

12 Stephanie Smith, *Feast: Radical Hospitality in Contemporary Art* (Chicago, IL: Smart Museum of Art, 2013), p. 129.

13 Rosler, *Decoys and Disruptions*, p. 7.

14 Margaret Morse, *Virtualities: Television, Media Art, and Cyberculture* (Bloomington, IN: Indiana University Press, 1998), pp. 134–5.
15 Sometimes the desire to be fed was made very literal, as with Barbara T. Smith's *Feed Me* (1973) and the Feminist Art Workers' *Heaven or Hell* (1979). In *Feed Me*, Smith stands naked in a room, surrounded by food. Smith 'met audience members one at a time in a secluded room. A recording repeated "Feed Me"'. Montano, 'Food & art', p. 77. The Feminist Art Workers took over a dormitory cafeteria during lunch hour, where they 'performed two tableaux (heaven and hell) and placed 200 placemats and four-foot long forks on the tables so that others could join in (which they did)'. Montano, 'Food & art', p. 64. Both of these projects, Smith's especially, illustrate how the topic of food compels broader, more generous sensorial readings. Smith conjures a proprioceptive set of relations that activate external and internal haptic interactions.
16 Morse, *Virtualities*, p. 128.
17 Pandora Tabatabai Asbaghi and Jerry Gorovoy, *Louise Bourgeois: Blue Days and Pink Days* (Milan: Fondazione Prada, 1997), p. 98.
18 Chicago, *The Dinner Party*, pp. 13, 21–2.
19 Morse, *Virtualities*, p. 128.
20 Brooklyn Museum, 'Brooklyn Museum: Elizabeth A. Sackler Center for Feminist Art: *The Dinner Party*', n.d. Available at http://www.brooklynmuseum.org/eascfa/dinner_party/ (accessed 10 February 2015).
21 Chicago, *The Dinner Party*, p. 11.
22 Linda Montano, *Performance Artists Talking in the Eighties* (Berkeley, CA: University of California, 2000), pp. 179–180.
23 Norma Broude and Mary D. Garrard (eds), *The Power of Feminist Art* (New York: Harry N. Abrams, 1994), p. 235.
24 Suzanne Lacy (ed.), *Mapping the Terrain: New Genre Public Art* (Seattle, WA: Bay Press, 1995), pp. 251–3.
25 'Lucy's schedule', written by Jess Oppenheimer, Madelyn Pugh-Davis and Bob Carroll, Jr with Lucile Ball, Desi Arnaz, and Vivian Vance, *I Love Lucy* (CBS, 26 May 1952).
26 Montano, *Performance Artists Talking in the Eighties*, p. 151.

5 DIY Coke

1 Martha Rosler, 'Romances of the meal', in Hans Ulrich Obrist and Barbara Vanderlinden, *Laboratorium* (Antwerp: Dumont, 1999), p. 259.
2 For an exploration of Coca-Cola as a cooking ingredient, see Anne E. McBride, 'Have your Coke and eat it too: what cooking with Coca-Cola says about cultural imperialism', *Gastronomica: The Journal of Food and Culture* 5/1 (2005), pp. 80–7.

3 Obrist and Vanderlinden, *Laboratorium*, p. 249.
4 See Catherine Beecher, *A Treatise on Domestic Economy: For the Use of Young Ladies at Home and at School* (Boston, MA: Thomas Webb, 1843) and Christine Frederick, *Household Engineering: Scientific Management in the Home* (Chicago, IL: American School of Home Economics, 1920).
5 Christian Boltanski, Bertrand Lavier and Hans Ulrich Obrist, 'DO-IT at e-flux', e-flux.com. Available at http://www.e-flux.com/projects/do_it/homepage/do_it_home.html (accessed 30 January 2014).
6 Hans Ulrich Obrist and Kate Fowlie (eds), *Do It: The Compendium* (New York: Distributed Art Publishers, 2013).
7 'Original recipe', *This American Life*, National Public Radio, WNYC, New York, 11 February 2011.
8 Mark Pendergrast, *For God, Country and Coca-Cola: The Definitive History of the Great American Soft Drink and the Company that Makes It* (New York: Basic Books, 1993).
9 To support the importance of the artist as researcher, e-flux also hosts a searchable database called 'Martha Rosler library', which reproduces Rosler's bookshelves as text and photographs. 'Martha Rosler library', e-flux. Available at http://www.e-flux.com/projects/library/ (accessed 30 January, 2014).
10 *King Corn*, dir. Aaron Woolf (Mosaic: 2007).
11 Beatriz da Costa, 'Reaching the limit: when art becomes science', in Beatriz da Costa and Kavita Philips (eds), *Tactical Biopolitics* (Cambridge, MA: MIT Press, 2008), pp. 365–82, p. 373.
12 Ibid., p. 384.
13 Beatriz da Costa, *Anti Cancer Survival Kit*. Available at http://www.rockethub.com/projects/12551-the-anti-cancer-survival-kit#description-tab (accessed 30 January 2014).
14 Robert Mitchell, *Bioart and the Vitality of Media* (Seattle, WA: Washington University Press, 2010), p. 64, and Hannah Higgins, *FLUXUS Experience* (Berkeley: University of California Press, 2002), p. 98.
15 Natalie Jeremijenko and Heath Bunting, *Biotech Hobbyist Magazine*, n.d. Available at http://www.nyu.edu/projects/xdesign/biotechhobbyist/skin.html (accessed 30 January 2014).
16 Eduardo Kac, *Cypher*, n.d. Available at http://www.ekac.org/cypher.text.html (accessed 30 January 2014).
17 Tissue Culture and Art Project, *DIY DVK*. Available at http://tcaproject.org/projects/victimless/dvk01 (accessed 30 January 2014) (link no longer working).
18 Natasha Lushetich, 'Hovering between being and non-being: Nishidian inter-expression and the Fluxkit', *Performance Research: A Journal of the Performing Arts* 16/4 (2011), pp. 73–9.
19 Ibid., p. 73.

20 Higgins, *FLUXUS Experience*, p. 98.
21 Robert Mitchell notes the overlap between labs and kitchens when asking how the laboratory might become more accessible: 'I remember being very struck by the extent to which the "genre" of lab protocols [...] resembled a "genre" with which I was much more familiar: namely, kitchen recipes.' Mitchell, *Bioart and the Vitality of Media*, p. 130.

6 Meat Culture

1 Mary C. Phelan, 'Techniques for mammalian cell tissue culture', *Current Protocols in Cytometry*, 36:A.3B.1–A.3B.18, 1 May 2006. Available at http://onlinelibrary.wiley.com/doi/10.1002/0471142956.cya03bs36/full#cya03b-rec-0001 (accessed 30 January 2014; emphasis in original).
2 See Oron Catts and Gary Cass's reflection on leading SymbioticA workshops, 'Labs shut open: a biotech hands-on workshop for artists', in Beatriz da Costa and Kavita Philips (eds), *Tactical Biopolitics: Art, Activism, and Technoscience* (Cambridge, MA: MIT Press, 2008), pp. 143–56, p. 147.
3 Hannah Landecker, *Culturing Life: How Cells Became Technologies* (Cambridge: Harvard University Press, 2007), p. 1.
4 Christopher Henke and Thomas Gieryn, 'Sites of scientific practice: the enduring importance of place', in Edward Hackett, Olga Amsterdamska, Michael Lynch and Judy Wajcman (eds), *The Handbook of Science and Technology Studies*, 3rd edn (Cambridge, MA: MIT Press, 2008), pp. 353–76, p. 360.
5 Verena Pereval addresses this issue in architecture in her paper 'Architecture of a demonstration: building facts, walls and publics', 4S & EASST Conference, 2004. Abstract available at http://www.csi.ensmp.fr/WebCSI/4S/abstract/abstract_papier.php?numero=36 (accessed 30 January 2014).
6 Nicola Triscott, 'Performative science: the case of Critical Art Ensemble', in Maria Chatzichristodoulou, Janis Jeffries and Rachel Zerihan (eds), *Interfaces of Performance* (London: Ashgate, 2009), pp. 153–66, p. 157.
7 Draft addition available at OED website, http://www.oed.com/view/Entry/60835?redirectedFrom=embed#eid (accessed 30 January 2014).
8 Kylie Tuosto, 'The "grunt truth" of embedded journalism: the new media/military relationship', *Stanford Journal of International Relations* 20 (Fall/Winter 2008), pp. 20–31.
9 The Tissue Culture and Art Project, 'TC&A biographies of Project members'. Available at http://www.tca.uwa.edu.au/atGlance/bios.html (accessed 30 January 2014).
10 Taking the embedded collaboration as a model for how artists and scientists would interact at SymbioticA, the laboratory occupies the second floor of the School of Anatomy and Human Biology at The University of Western Australia.

At any given time SymbioticA hosts artists, scholars and scientists producing work that could only be made in a fully equipped biology laboratory. Recent work at SymbioticA includes a series of death masks of laboratory mice by Verena Kaminiarz, experimental films about internal processes by Gail Wight and Orlan's *Harlequin Coat*: 'a biotechnological coat, consisting of *in vitro* skins in coloured diamond shaped petri dishes [...] symbolis[ing] cultural crossbreeding'. Like Tissue Culture and Art Project's residency at the Tissue Engineering and Organ Fabrication Lab, these projects situate artists inside laboratory environments, in conversation with scientists, but SymbioticA focuses on producing art. At Harvard, art was one of several research foci, but at UWA participating scientists usually have their own laboratories. (For example, Stuart Bunt's neurovascular research does not happen at the SymbioticA laboratory benches, although 'many of the scientists working in SymbioticA say they get challenges, both intellectual and practical, from working with the artists'). Stuart Bunt, 'The role of the scientist and science in bio-art', in Melentie Pandilovski (ed.), *Art in the Biotech Era* (Adelaide: Experimental Art Foundation, 2008), pp. 62–7, p. 63.

11 Oron Catts and Gary Cass, 'Biotech art workshop conducted by Symbiotica', presented at the conference *Bio-Art and the Public Sphere*, University of California, Irvine, 10–14 October 2005.

12 Oron Catts and Ionat Zurr, 'Semi living art', in Eduardo Kac (ed.), *Signs of Life* (Cambridge, MA: MIT Press, 2007), p. 232.

13 The Laboratory for Tissue Engineering and Organ Fabrication. Available at http://www.massgeneral.org/research/researchlab.aspx?id=1129 (accessed 30 January 2014).

14 Tissue Engineering and Organ Fabrication Laboratory. Available at http://web.archive.org/web/20041031170118/www.massgeneral.org/tissue/projects/mains/amain.html (accessed 30 January 2014).

15 Hannah Landecker, 'Living differently in time: plasticity, temporality, and cellular biotechnologies', *Culture Machine* 7 (2005). Available at http://www.culturemachine.net/index.php/cm/article/view/26/33 (accessed 30 January 2014).

16 Ibid.

17 Steve Baker, 'Sloughing the human', in Cary Wolfe (ed.), *Zoontologies* (Minneapolis, MN: Minnesota University Press, 2003), pp. 147–64.

18 Catts and Zurr, 'Semi living art', p. 233.

19 Monica Bakke, 'Zoe-philic desires: wet media art and beyond', *parallax* 14/3 (2008), pp. 21–34, p. 22.

20 Ibid.; Robert Mitchell, *Bioart and the Vitality of Media* (Seattle, WA: Washington University Press, 2010), p. 27.

21 Jens Hauser, *L'art biotech* (Nantes: Le Lieu Unique, 2003).

22 Ionat Zurr and Oron Catts, 'The ethical claims of bioart: killing the other or self cannibalism?', *Australia and New Zealand Journal of Art: Art and Ethics* 4/2 (2003) and 5/1 (2004), pp. 167–88.

23 See 'Ross G. Harrison' in Donna Haraway, *Crystals, Fabrics, and Fields: Metaphors That Shape Embryos* (Berkeley, CA: North Atlantic, 2003), pp. 64–100, and Theodore Malinin, *Surgery and Life: The Extraordinary Career of Alexis Carrel* (New York: Harcourt Brace Jovanovich, 1979), pp. 55–60.

24 Tissue Culture and Art Project, 'Semi-living food – "disembodied cuisine"'. Available at http://www.tca.uwa.edu.au/disembodied/dis_text.html (accessed 30 January 2014).

25 Catts and Zurr, 'Semi living art', p. 244.

26 Donna Haraway, *When Species Meet* (Minneapolis, MN: Minnesota University Press, 2008), p. 296.

27 Catts and Zurr, 'Semi living art', p. 233.

28 Oron Catts and Ionat Zurr, 'Towards a new class of being: the extended body', *Artnodes* 6 (2006). Available at http://www.uoc.edu/artnodes/6/dt/eng/catts_zurr.pdf (accessed 1 May 2015).

29 For more on ecocritique, see Timothy Morton, *Ecology Without Nature: Rethinking Environmental Aesthetics* (Cambridge, MA: Harvard University Press, 2007). For a central ecofeminist text, see Carolyn Merchant, *Earthcare: Women and the Environment* (New York: Routledge, 1995).

30 Jacob Metcalf, 'Meet shmeat: food system ethics, biotechnology and re-worlding technoscience', *parallax* 19/1 (2013), pp. 74–87, p. 84 (emphasis in original).

31 Complex in its commonalities, certain figures are privileged in the contact zones presented by *Disembodied Cuisine*. Human farmers and factory workers are not represented, while human cooks and consumers are privileged. Tissue culturists replace farmers and factory workers, and tissue culturists are elite and few in number. This framing of the project makes it difficult to address questions of labour or to intervene in current methods of meat production like slaughterhouse culture and factory farming. Recent work, including *Futile Labor* (2015), addresses these missing actors and provokes engagement with these issues.

32 Catts and Zurr, 'Semi living art', p. 242.

33 W.F. van Eelen, W.J. van Kooten and W. Westerhof hold a Dutch patent on scaffold technologies called 'Industrial scale production of meat from *in vitro* cell cultures'. Cited in P.D. Edelman, D.C. McFarland, V.A. Mironov, J.G. Matheny, '*In vitro*-cultured meat production', *Tissue Engineering* 11/5–6 (2005), pp. 659–62, p. 662.

34 'The first laboratory that tried to produce meat *in vitro* was SymbioticA, a laboratory of the University of Western Australia, where artists work together with tissue engineers. First they tried to produce some meat from the muscle stem cells of sheep. That didn't work out so well. But later they indeed succeeded to grow muscle tissue from a frog. They presented the rather minuscule frog meat with a calvados sauce on a exhibition in Nantes, France.' Marianne Heselmans, 'Cultivated meat (NRC)', *New Harvest*, 10 September

2005. Available at http://www.new-harvest.org/2005/09/cultivated-meat-nrc/ (accessed 30 January 2014).

35 Edelman et al., '*In vitro*-cultured meat production', p. 659.

36 Mitchell, *Bioart and the Vitality of Media*, p. 27.

37 Jens Hauser, 'Bio art – taxonomy of an etymological monster', *Hybrid: Living in Paradise* (Ars Electronica, 2005). Available at http://90.146.8.18/en/archives/festival_archive/festival_catalogs/festival_artikel.asp?iProjectID=13286 (accessed 30 January 2014).

38 Laura Mulvey, 'Visual pleasure and narrative cinema', *Screen* 16/3 (1975), pp. 6–18.

39 Rachel Carson, *Silent Spring* (New York: Houghton Mifflin, 2002), pp. 205–6.

40 Catts and Zurr, 'Ethical claims', p. 13.

41 Malinin, *Surgery and Life*, p. 55. See also Haraway, *Crystals, Fabrics, and Fields*; Susan Squier, 'Life and death at Strangeways: the tissue culture point of view', in Paul Brodwin (ed.), *Biotechnology and Culture: Bodies, Anxieties, Ethics* (Bloomington, IN: Indiana University Press, 2000), pp. 27–52, p. 28 and Hannah Landecker, 'Immortality, *in vitro*: a history of the HeLa cell line', ibid., pp. 53–74, p. 55, and Landecker, *Culturing Life*, pp. 28–67.

42 Alexis Carrel and Charles Lindbergh, *The Culture of Organs* (New York: Hoeber, 1938), p. 7.

43 Haraway, *Crystals, Fabrics, and Fields*, p. 42.

44 Ionat Zurr, 'Growing semi-living art', PhD thesis, University of Western Australia 2009, p. 145. Available at http://theses.library.uwa.edu.au/adt-WU2009.0125/ (accessed 29 January 2014).

45 Often research budgets, public relations and commercial outcomes collide in ways that support Zurr's point that animal rights organisations fail to understand the ethical costs of *in vitro* meat. For example, in 2008, People for the Ethical Treatment of Animals (PETA) offered a symbolic $1 million prize to researchers capable of producing *in vitro* chicken at an industrial scale by 2014. As the deadline for the contest approached in 2014, PETA reconsidered the prize in light of advances with *in vitro* beef and hamburger, making the vague promise that 'to put that $1 million prize (originally intended for the creator of in vitro chicken) to good use in combating cruelty in food production in other ways'. 'PETA's "In vitro" chicken contest', n.d. Available at http://www.peta.org/features/vitro-meat-contest/ (accessed 18 April 2014). Also see John Schwartz, 'PETA's latest tactic: $1 million for fake meat', *New York Times*, 21 April 2008. Available at http://www.nytimes.com/2008/04/21/us/21meat.html (accessed 29 January 2014). Also see press release, PETA, 'In vitro meat prize deadline extended', 25 June 2012. Available at http://www.peta.org/blog/vitro-meat-prize-deadline-extended/ (accessed 29 January 2014).

46 Edelman et al., '*In vitro*-cultured meat production', p. 660.

47 Oron Catts and Ionat Zurr, 'Disembodied livestock: the promise of a semi-living utopia', *parallax* 19/1 (2013), pp. 101–13.

48 Edelman et al., '*In vitro*-cultured meat production', p. 661.
49 PETA, 'About PETA'. Available at http://www.peta.org/about-peta/ (accessed 7 February 2014).
50 Zurr and Catts, 'Ethical claims', pp. 13–4.
51 Ionat Zurr elaborates on this point in her dissertation; see excerpt above: Zurr, 'Growing semi-living art', p. 167.
52 Catts and Cass, 'Biotech art workshop'.
53 Carol Gigliotti, 'Leonardo's choice: the ethics of artists working with genetic technologies', *AI and Society* 20 (2006), pp. 22–34, p. 33.
54 Ibid., p. 33.
55 Although Gigliotti uses the word 'transgenic' to signal this complex system of interdependence, a much broader set of practices including agriculture, horticulture and tissue culture, which are not necessarily transgenic practices, must be considered alongside genetic manipulation.
56 Zurr, 'Growing semi-living art', p. 167.
57 Park Doing, 'Give me a laboratory and I will raise a discipline: the past, present, and future politics of laboratory studies in STS', in Hackett et al., *Handbook of Science and Technology Studies*, pp. 279–98, p. 291.
58 Bruno Latour, 'Give me a laboratory and I will raise the world', in Karin Knorr-Cetina and Michael Mulkay (eds), *Science Observed: Perspectives on the Social Study of Science* (Beverly Hills, CA: Sage, 1983), pp. 141–70, p. 160.

7 Public Amateurism

1 Critical Art Ensemble, Beatriz da Costa and Shyh-shiun Shyu, *Free Range Grain*, n.d. Available at http://critical-art.net/Original/free/freeRange.swf (accessed 4 February 2014).
2 Robert Mitchell, *Bioart and the Vitality of Media* (Seattle, WA: Washington University Press, 2010), p. 36.
3 For more about CAE's move into biotechnology, see Critical Art Ensemble, Jon McKenzie and Rebecca Schneider, 'Critical Art Ensemble: tactical media practitioners', *The Drama Review* 44/4 (Winter 2000), pp. 136–50.
4 Critical Art Ensemble, *Digital Resistance: Explorations in Tactical Media* (Brooklyn, NY: Autonomedia, 2001), p. 59.
5 Eugene Thacker, 'Data made flesh: biotechnology and the discourse of the posthuman', *Cultural Critique* 53 (Winter 2003), pp. 72–97.
6 See CAE performances *Flesh Machine* (1997–8) and *Cult of the New Eve* (1999–2000).
7 Beatriz da Costa, 'The art of biologies and environmentalism', University of California Santa Cruz Digital Art and New Media Festival symposium, 5 May 2006.

8 European Commission, 'Directive 2001/18/EC of the European Parliament on the deliberate release of genetically modified organisms', 12 March 2001. Available at http://eur-lex.europa.eu/LexUriServ/LexUriServ.do?uri=CELEX:32001L0018:EN:HTML (accessed 4 February 2014).

9 Critical Art Ensemble, da Costa and Shyu, *Free Range Grain*.

10 Ibid.

11 European Commission, 'Directive 2001/18/'.

12 Critical Art Ensemble, 'Fuzzy biological sabotage', 2013. Available at http://critical-art.net/MolecularInvasion.html (accessed 19 February 2015).

13 Claire Pentecost, 'What did you eat and when did you know it?', *Art Journal* 61/3 (Autumn 2002), pp. 47–62, p. 48.

14 Ron Arnold generally receives credit for this phrase, first published in his book, *Ecoterror: The Violent Agenda to Save Nature: The World of the Unabomber* (Bellevue, WA: Free Enterprise Press, 1997).

15 See Chapter 6, 'Meat Culture' for more about embedded art/science collaboration.

16 Claire Pentecost, quoted in Beatriz da Costa, 'Reaching the limit', in Beatriz da Costa and Kavita Philip (eds), *Tactical Biopolitics* (Cambridge, MA: MIT Press, 2008), p. 375.

17 Ibid.

18 Gregory Sholette, 'Disciplining the avant-garde: The United States versus the Critical Art Ensemble', *Circa* 112 (Summer 2005), pp. 50–9, p. 52.

19 Lucy Yau, 'Seized: Steve Kurtz and the art of dissent', *ArtVoice* 7/5 (2008). Available at http://artvoice.com/issues/v7n25/seized (accessed 19 February 2015).

20 Critical Art Ensemble is not the only group to be uninvited from bringing certain types of material to shows in the US. See, for example, Guillermo Gómez-Peña, 'Disclaimer', *The Drama Review* 50/1 (Spring 2006), pp. 149–58.

21 Schneider and McKenzie reproduce the sign left by CAE at the 'aborted exhibition' in 2004: 'While preparing *Free Range Grain* for The Interventionists exhibition, CAE/da Costa's lab equipment was seized by the FBI. Lawyers believe that the confiscation of the hardware and wetware could have been made possible through the use of the PATRIOT Act. Constitutional rights lawyers are currently considering the validity of this linkage. The lab equipment used for *Free Range Grain* could not be successfully used for the production and weaponization of any germs dangerous to humans or animals. Moreover, the FBI field and laboratory tests have demonstrated that the equipment seized was not used for any illegal purpose. Furthermore, any person in the U.S. May obtain and possess this equipment without violating U.S. laws. In spite of these facts, the FBI has persisted in its investigation of bioterrorism with respect to the research conducted in this project, and, as of this writing, the FBI has not released any of the seized items, leaving us

unable to present the project in its entirety.' Critical Art Ensemble, McKenzie and Schneider, 'Tactical media practitioners', pp. 5–6.

22 Nicola Triscott, 'Performative science in an age of specialization: the case of Critical Art Ensemble', in Maria Chatzichristodoulou, Janis Jefferies and Rachel Zerihan (eds), *Interfaces of Performance* (Farnham: Ashgate, 2009), pp. 153–66, p. 166.

23 Bronislaw Szerszynski, 'The post-ecologist condition: irony as symptom and cure', *Environmental Politics* 16/2 (2007), pp. 337–55, p. 346.

24 Ibid., pp. 347–8.

25 Oliver Watts, 'The image and the terrorist', *Law Text Culture* 10 (2006), pp. 221–38, p. 222.

26 Ibid., p. 226.

27 The Critical Art Ensemble Defense Fund website has gathered numerous statements of support from the art and science communities. See 'The Critical Art Ensemble Defense Fund'. Available at http://www.caedefensefund.org/ (accessed 19 February 2015).

28 *Strange Culture*, dir. Lynn Hershman-Leeson (L5 Productions, 2007).

29 Carolyn Thompson, 'Prof sent bacteria to artist; no jail', Associated Press, 12 January 2008. Available at http://www.usatoday.com/news/nation/2008-02-12-3296235700_x.htm (accessed 3 February 2014); Paula Reed Ward, 'Doctor guilty of mailing bacteria for use in artwork', *Pittsburgh Post-Gazette*, 12 October 2007. Available at http://www.post-gazette.com/pg/07285/824828-85.stm (accessed 3 February 2014).

30 George J. Annas, 'Bioterror and "bioart" – a plague o' both your houses', *The New England Journal of Medicine* 354/25 (2006), pp. 2715–20; Sholette, 'Disciplining the avant-garde'; Anna Munster, 'Why is bioart *not* terrorism?: some critical nodes in the networks of infomatice life', *Culture Machine* 7 (2005). Available at http://www.culturemachine.net/index.php/cm/article/view/31/38 (accessed 3 February 2014); Joyce Lok See Fu, 'The potential decline of artistic activity in the wake of the Patriot Act: the case surrounding Steven Kurtz and the Critical Art Ensemble', *The Columbia Journal of Law and the Arts* 29/1 (Fall 2005), pp. 83–106.

31 Munster, 'Why is bioart *not* terrorism?'.

32 Note the subtitle of former ELF spokesperson Craig Rosebraugh's memoir, *Burning Rage of a Dying Planet: Speaking for the Earth Liberation Front* (New York: Lantern Books, 2004).

33 Natalie Jeremijenko and Heath Bunting, *Biotech Hobbyist Magazine*, n.d. Available at http://www.nyu.edu/projects/xdesign/biotechhobbyist/bio_about.html (accessed 18 February 2015). The projects featured on the website include 'Tree cloning' and 'Skin culture'.

34 Critical Art Ensemble, McKenzie and Schneider, 'Tactical media practitioners', p. 9 (emphasis in original).

35 Mitchell, *Bioart and the Vitality of Media*, pp. 48–51.

36 Munster, 'Why is bioart *not* terrorism?'

37 Stephanie Kane and Pauline Greenhill, 'A feminist perspective on bioterror: from anthrax to the critical art ensemble', *Signs: Journal of Women in Culture and Society* 33/1 (2007), pp. 53–80.

38 *Strange Culture*, Hershman-Leeson (2007).

39 Timothy Stock and Warren Heise, 'Suspect culture: an exposition of the death of Hope Kurtz', in John Knechtel (ed.), *Suspect* (Cambridge, MA: MIT Press, 2006), pp. 332–49.

40 For more about Butler and Boisvert, see Annas, 'Bioterror and "bioart"'. Butler was sentenced to two years in prison and over $50,000 in fines after being charged with smuggling bacteria samples into the United States, improperly transporting them within the country, theft, embezzlement and fraud. Boisvert installed black boxes printed with the word 'fear' around New York City subway stations, and was charged with reckless endangerment. Star Simpson was detained at Boston's Logan International Airport after she was spotted by authorities wearing a breadboard mounted on a sweatshirt and was charged with possession of a hoax device, later reduced to disorderly conduct. See Simpson's interview on Xeni Jardin, 'BBtv – Star Simpson's fuzzy logic, MacGyver, MIT lasers, and trippy glasses: Maker Faire with Phil Torrone', 8 May 2008. Available at http://www.boingboing.net/2008/05/08/bbtv-star-simpsons-f.html (accessed 4 February 2014).

41 Following the release of *Strange Culture* and the dismissal of charges against Kurtz, Critical Art Ensemble mounted an exhibition at Hallwalls Contemporary Arts Center in Buffalo, New York. In collaboration with the Institute for Applied Autonomy, CAE presented 'Seized', which displayed the garbage left behind after the Joint Terrorism Task Force raided Kurtz's home in Buffalo. Despite the overflowing containers filled with Gatorade bottles, hazmat suits, discarded packaging and other detritus, 'Seized' is an exhibition about absence. The artwork, computers, books and manuscripts the Joint Terrorism Task Force took away are revealed through photographs of the empty places in Kurtz's house and documentation and ephemera from the performances interrupted by the raid. Originally planned to coincide with Kurtz's trial, 'Seized' may have been a catalyst for the return of Kurtz's possessions; the show opened on 8 June 2008, and the CAE Defense Fund announced that 'the FBI has finally returned the art projects, research materials and personal belongings seized in its 2004 raid on Steve's home' on 11 June 2008. Both 'Seized' and *Strange Culture* focus on materiality and the importance of context, revealing the complex ways everything from art to garbage travels between public and private spaces. The circulation of *S. marcescens*, pizza boxes, books and computers can be considered

alongside the circulation of knowledge and ideas. CAE's practice depends on critically connecting materials with ideas in spaces like the art laboratory and the art studio. See 'The Critical Art Ensemble Defense Fund' and 'Hallwalls Contemporary Arts Center: Visual Arts'. Available at http://www.hallwalls.org/visual_shows/2008/show_seized.html (accessed 4 February 2014).

8 Cookbook

1 Christine Chin, *The Genetically-Modified Foods Cook Book* (2004). Available at http://gmcookbook.com/ (accessed 18 March 2014).

2 The rhetoric of these books varies, sometimes focusing on health and nutrition, sometimes veering into paranoid anti-science polemic. For a range of examples, see Courtney Pineau, *The Non-GMO Cookbook: Recipes and Advice for a Non-GMO Lifestyle* (New York: Skyhorse, 2013); Mae-Wan Ho and Lim Li Ching, *GMO Free: Exposing the Hazards of Biotechnology to Ensure the Integrity of Our Food Supply* (Ridgefield, CT: Vital Health, 2004); Matthew Johnson, *GMO Free Diet: How to Stay Healthy by Identifying and Avoiding Dangerous Foods* (Digital Direct Publishing, 2013); and Wendy Cottiers-Pacella, *The Bare Truth: Living GMO-Free in the 21st Century* (Fort Lauderdale, FL: Positive Nutrition, 2014).

3 Many of the 'genetically modified' ingredients in Stadler's cookbook are varieties of plants that farmers have been selectively breeding over time. Most activists and GMO-free cookbook authors focus on recombinant DNA technologies rather than the long history of selective breeding. I have not managed to obtain a copy of Stadler's cookbook, but an order form is available at its pro-GMO lobbying publisher's webpage, http://www.internutrition.ch/in-news/form.html (accessed 18 March 2014). The book has appeared in many publications, including *Nature Genetics*: 'News and Views', *Nature Genetics* 29, 249 (2001). Beda Stadler, *Gene an die Gabel* (Zurich: Internutrition, 2001).

4 Jorjo Riss (ed.), *Good Food Gone Bad: Genetically Modified Recipes for Disaster* (Brussels: Greenpeace, 2010). Available at http://www.greenpeace.org/eu-unit/en/Publications/2010/GMO-cookbook/ and http://www.greenpeace.org/eu-unit/Global/eu-unit/reports-briefings/2010/4/GMO-cookbook.pdf (accessed 18 January, 2014).

5 For an example of this return to recipes from the 1980s, see Lisa Heldke, 'Recipes for theory making', *Hypatia* 3/2 (Summer 1988), pp. 15–29. See also Arlene Voski Avakian and Barbara Haber (eds), *From Betty Crocker to Feminist Food Studies: Critical Perspectives on Women and Food* (Boston, MA: Massachusetts University Press, 2005), Janet Theophano, *Eat My Words: Reading Women's Lives through the Cookbooks They Wrote* (New York: Palgrave, 2002), Joanne Hollows, 'The feminist and the cook: Julia Child, Betty Friedan and Domestic Femininity', in Lydia Martens and Emma Casey (eds), *Gender and Consumption: Domestic*

Cultures and the Commercialisation of Everyday Life (Farnham: Ashgate, 2007), pp. 33–48 and Jennifer R. Horner, '*Betty Crocker's Picture Cookbook*: a gendered ritual response to social crises of the postwar era', *Journal of Communication Inquiry* 24 (2000), pp. 332–45.

6 Christine Chin, *Vegetable Human Hybrids*. Available at http://christinechin.net/hybrids/index.html (accessed 18 March 2014).

7 Christine Chin, personal correspondence, 12 October 2006.

8 Chin, *The Genetically Modified Foods Cook Book*.

9 Christine Chin, personal correspondence, 12 October 2006.

10 Ibid.

11 Lewis Petrinovich, *The Cannibal Within* (New York: Aldine de Gruyter, 2000), p. 193.

12 Priscilla Walton, *Our Cannibals, Ourselves* (Urbana, IL: Illinois University Press, 2004), p. 49.

13 Ibid., p. 110.

14 Ibid., p. 199; Michael Pollan, *The Omnivore's Dilemma: A Natural History of Four Meals* (New York: Penguin, 2006), pp. 75–6.

15 Ibid., p. 76.

16 Ray Tannahill, *Flesh and Blood: The History of the Cannibal Complex* (Boston, MA: Little, Brown, 1996), p. viii.

17 Petrinovich, *The Cannibal Within*, p. vii.

18 Donna Haraway, *Modest_Witness@Second_Millennium.FemaleMan©_Meets_OncoMouse: Feminism and Technoscience* (New York: Routledge, 1997), p. 60.

19 Next Nature Network, *The In Vitro Meat Cookbook* (Amsterdam: BIS Publishers, 2014), p. 7.

20 Ibid., pp. 145, 147.

21 Isha Datar and Robert Bolton, 'The carnery', in Next Nature Network, *In Vitro Meat Cookbook*, p. 154.

22 Next Nature Network, *In Vitro Meat Cookbook*, p. 55.

23 Jacob Metcalf, 'Meet shmeat: food system ethics, biotechnology and re-worlding technoscience', *parallax* 19/1 (2013), pp. 74–87.

24 Ibid., p. 83 (emphasis in original).

25 Ibid., pp. 83–2.

26 Critical Art Ensemble, 'Betty Crocker 3000 presents food for a hungry world', in *Molecular Invasion* (Brooklyn, NY: Autonomedia, 2002), p. 131. See Chapter 9, 'Carnal Light', for a more detailed reading of one of the *Betty Crocker 3000* recipes.

27 Ibid., p. 131.

28 The Grocery Manufacturer's Association has petitioned the US Food and Drug Administration to regulate and define the word 'natural'. The agency currently declares that 'from a food science perspective, it is difficult to define a food product that is "natural" because the food has probably been processed and is

no longer the product of the earth'. See Stephanie Strom, 'Group seeks special label for food: "natural"', *New York Times*, 19 December 2013. Available at http://www.nytimes.com/2013/12/20/business/trade-group-seeks-natural-label-on-modified-food.html?_r=0 (accessed 24 April 2014). See also US Food and Drug Administration (FDA), 'FDA Basics: What is the meaning of "natural" on the label of food?', n.d. Available at http://www.fda.gov/aboutfda/transparency/basics/ucm214868.htm (accessed 24 April 2014).

29 Betty Crocker, *Betty Crocker's Picture Cookbook* (Minneapolis: Wiley, 1950), p. 35.

30 Laura Shapiro, '"I guarantee": Betty Crocker and the woman in the kitchen', in Avakian and Haber, *From Betty Crocker to Feminist Food Studies*, pp. 29–40, p. 29.

31 Ibid., p. 30.

32 See 'Pel-Freez'. Available at http://www.pel-freez.com/ (accessed 24 April 2014).

9 Carnal Light

1 Critical Art Ensemble, 'Betty Crocker 3000 presents food for a hungry world', *Molecular Invasion* (Brooklyn, NY: Autonomedia, 2002), pp. 135–6.

2 Jacques Derrida, 'Force and signification', in *Writing and Difference*, trans. Alan Bass (Chicago, IL: Chicago University Press, 1980), p. 27.

3 Vivienne Baille Gerritsen, 'The greenest of us all?', *Protein Spotlight* 11 (June 2001). Available at http://web.expasy.org/spotlight/pdf/sptlt011.pdf (accessed 15 September 2012).

4 Bijal P. Trivedi, 'Introducing ANDi: The first genetically modified monkey', *Genome News Network*, 16 January 2001. Available at http://www.genomenewsnetwork.org/articles/01_01/ANDi.shtml (accessed 15 September 2012).

5 David Whitehouse, 'GM fish glows in the bowl', *BBC News*, 27 June 2003. Available at http://news.bbc.co.uk/2/hi/science/nature/3026104.stm (accessed 15 September 2012). The Center for Genomic Gastronomy has been inspired by GloFish to create 'Glowing Sushi', a recipe that 'uses everyday ingredients and some simple kitchen chemistry to explore cutting edge biotechnology'. Center for Genomic Gastronomy, 'Glowing sushi', n.d. Available at http://www.glowingsushi.com/ (accessed 15 September 2012).

6 Pimprapar Wongsrikeao, Dyana Saenz, Tommy Rinkoski, Takeshige Otoi and Eric Poeschla, 'Antiviral restriction factor transgenesis in the domestic cat', *Nature Methods* 8 (2011), pp. 853–9. Available at http://www.nature.com/nmeth/journal/v8/n10/full/nmeth.1703.html (accessed 15 September 2012).

7 F.S. Hsiao, W.S. Lian, S.P. Lin, C.J. Lin, Y.S. Lin et. al., 'Toward an ideal animal model to trace donor cell fates after stem cell therapy: production of stably labeled multipotent mesenchymal stem cells from bone marrow of

transgenic pigs harboring enhanced green fluorescence protein gene', 24 June 2011. Available at http://www.ncbi.nlm.nih.gov/pubmed/21705633 (accessed 15 September 2012).

8 GFP's visual impact makes it a popular topic in interdisciplinary art and science education, with many students initially working with bacteria. For example, see Jennifer Willet, 'Part 1: GFP LAB', *Bioart: Contemporary Art and the Life Sciences*, 13 February 2012. Available at http://bioartwindsor.blogspot.com/2012/02/part-1-gfp-lab.html (accessed 15 September 2012); William Ward, 'Experiments with GFP – the art and the science', n.d. Available at http://dbm.rutgers.edu/11115433SpecialTopicsBiochem.html (accessed 30 January 2015); SymbioticA, 'BioTech Art Workshop' workshop programme, University of California Irvine. Available at http://www.symbiotica.uwa.edu.au/activities/workshops/SymbioticA_BioTech_art_workshop (accessed 15 September 2012).

9 George Gessert, *Green Light: Toward an Art of Evolution* (Cambridge, MA: MIT Press, 2010).

10 An early version of this chapter was given as a talk on the panel 'Tranimals' at 'Zoontotechnics: animality/technicity,' 14 May 2010. A further elaboration on 'tranimals' was published as 'Tranimals', *Transgender Studies Quarterly* 1/1–2 (May 2014), pp. 226–8.

11 A connection such that elements 'form an ensemble, a physical or metaphysical whole, the existence or idea of one being included in the existence or idea of another'. Paul Ricoeur, *Rule of Metaphor: The Creation of Meaning in Language* (New York: Routledge, 2003), p. 56.

12 'Metaplasm means a change in a word, for example by adding, omitting or transposing its letters, syllables or sounds [...] There is a biological taste to "metaplasm" [...] Flesh and signifier, bodies and words, stories and worlds: these are joined in naturecultures. Metaplasm can signify a mistake, a stumbling, a troping that makes a fleshly difference.' Donna Haraway, *The Companion Species Manifesto: Dogs, People and Significant Otherness* (Chicago, IL: Prickly Paradigm Press, 2003), pp. 20–1.

13 As a continuation of the *GFP Bunny* project, Kac transmitted 'lagoglyphs' into outer space. The *Lagoglyphs* are a rabbitographic form of language evoking dissection organs and bodily interiors; these compositions create 'a visual language that alludes to meaning but resists interpretation'. The transmission was accomplished through satellite broadcasting equipment and a parabolic dish antenna. This art action translates Alba into frequency, power, bandwidth and polarity. Art, here, is vibrational, rhythmic. It is in this way that we want to suggest that Alba is resonant, dissonant and improvisational. Available at http://www.ekac.org/gfpbunny.html (accessed 15 September 2012).

14 Definitions excerpted from the Oxford English Dictionary entries for 'drift, *v.*' and 'drift, *n.*' Available at http://www.oed.com/search?searchType=dictionary&q=drift&_searchBtn=Search (accessed 4 May 2015).

15 Dorion Sagan, 'Introduction: umwelt after Uexküll', in Jakob von Uexküll, *A Foray into the Worlds of Animals and Humans: With A Theory of Meaning* (Minneapolis, MN: Minnesota University Press, 2010), pp. 1–34, p. 1.

16 See Nicole Shukin, *Animal Capital: Rendering Life in Biopolitical Times* (Minneapolis, MN: Minnesota University Press, 2009).

17 Alphonso Lingis, *Phenomenological Explanations* (Dordrect: Martinus Nijhoff, 1986), p. 56.

18 Ibid., p. 67.

19 'Art enables matter to become expressive, to not just satisfy but also to intensify – to resonate and become more than itself [...] Art is the regulation and organization of its materials [...] according to self-imposed constraints, the creation of forms through which these materials come to generate and intensify sensation and this directly impact living bodies organs, nervous systems.' Elizabeth Grosz, *Chaos, Territory, Art: Deleuze and the Framing of the Earth* (New York: Columbia University Press, 2008), p. 4.

20 Ibid., p. 23.

21 Eva Hayward, 'FingeryEyes: Impressions of cup corals,' *Cultural Anthropology* 25/4 (2010), pp. 577–99.

22 The oldest and best example of this institutionally sanctioned art laboratory may be found at the University of Western Australia. 'SymbioticA.' Available at http://www.symbiotica.uwa.edu.au/ (accessed 15 September 2012).

23 Kristen Philipkoski, 'RIP: Alba, the glowing bunny', *WIRED*, 12 August 2002. Available at http://www.wired.com/medtech/health/news/2002/08/54399 (accessed 15 September 2012).

24 See Kathy High on rats as laboratory animals and as pets: *Embracing Animal* (2004). Available at http://www.embracinganimal.com/ (accessed 15 September 2012).

25 Eduardo Kac, *Telepresence and Bio Art: Networking Humans, Rabbits and Robots* (Ann Arbor, MI: Michigan University Press, 2005), p. 266.

26 Suzanne Anker, *The Molecular Gaze: Art in the Genetic Age* (Cold Spring Harbor, NY: Cold Spring Harbor Laboratory Press, 2003), p.108.

27 Rabbits prefer not to swim. One of the better known swimming rabbit news stories – US President Jimmy Carter's 1979 'rabbit attack' while fishing – involved a rabbit swimming as a last resort, escaping from hunting dogs. Carter's press secretary described the rabbit as 'a large, wet animal, making strange hissing noises and gnashing its teeth [...] intent upon climbing into the Presidential boat'. Jody Powell, *The Other Side of the Story* (New York: Morrow, 1984), p. 105.

28 Renova details further uses for their GFP rabbits: 'The specific relevant or derived GFP rabbit cell types can be used for gene therapy in animal models. GFP transgenic rabbits can also be used for deriving embryonic stem cells (from the early embryos), induced pluripotent stem (iPS) cells and further differentiation into many specific cell types.' Renova Life Inc., 'Products and services|GFP

rabbits', 15 January 2014. Available at http://www.renovalife.com/gfp_rabbits.php (accessed 15 September 2012).

29 Relatedly, the organisation New Life Animal Sanctuary is 'exclusively devoted to the rescue and rehabilitation of animals saved from laboratories' (including several large white albino bunnies like Alba); founder Gina Lynn was arrested blocking the delivery of rabbits to an undergraduate classroom at UC Santa Barbara before turning to sanctuary and adoption work. Rabbit Welfare Association and Fund, 'Breeds: choosing your perfect pet rabbit', n.d. Available at http://www.rabbitwelfare.co.uk/resources/?section=breeds.html (accessed 15 September 2012).

30 Eduardo Kac, *GFP Bunny*, n.d., nn. 17 and 18. Available at http://www.ekac.org/gfpbunny.html#gfpbunnyanchor (accessed 15 September 2012).

31 Alba's birth resulted from an established strategy for creating GFP rabbits: 'direct microinjection of DNA into the pronucleus of a rabbit zygote'. Kac, *Telepresence*, p. 264.

32 A recent episode of the BBC television series *Sherlock* featured a laboratory technician mother who brought a GFP bunny home as a pet for her daughter.

33 Critical Art Ensemble, 'Betty Crocker 3000 presents food for a hungry world', pp. 135–136.

34 Trivedi, 'Introducing ANDi'.

35 Donna Haraway, *When Species Meet* (Minneapolis, MN: Minnesota University Press, 2008), p. 17.

36 Robert Foley, 'Are we the only animals to cook our food?', 16 November 2008. Available at http://www.thenakedscientists.com/HTML/content/latest-questions/question/2169/ (accessed 15 September 2012).

37 Julie Ann Smith, 'Beyond dominance and affection: living with rabbits in post-humanist households', *Society and Animals* 11/2 (2003), pp. 181–97, p. 187.

38 Kim Severson, 'Don't tell the kids', 3 March 2010. Available at http://query.nytimes.com/gst/fullpage.html?res=9905E1D8103AF930A35750C0A9669D8B63&pagewanted=all (accessed 15 September 2012).

39 Stories were gathered from an informal poll on Facebook.

40 Julie Ann Smith, 'Beyond dominance and affection', pp. 187–8.

41 Donna Haraway, *When Species Meet*, p. 19.

42 Jonathan M. Gitlin, 'Is this sausage supposed to be green?', 12 January 2006. Available at http://arstechnica.com/science/news/2006/01/2477.ars (accessed 15 September 2012).

43 Cary Wolfe, 'From dead meat to glow in the dark bunnies: seeing "the animal question" in contemporary art', *parallax* 12/1 (2006), pp. 95–109, p. 107 (emphasis in original).

44 Renova Life, Inc.

45 Erik Stokstad, 'Engineered fish: friend or foe of the environment?' *Science* 297/5588 (13 September 2002), pp. 1797–9, p. 1797.

46 Francis Bartkowski, *Kissing Cousins: A New Kinship Bestiary* (New York: Columbia University Press, 2008), p. 157.

47 This chapter was first published in as 'Carnal light: following the white rabbit', *parallax* 66, 'bon appetit' (January 2013), pp. 114–127.

10 From Sanitation to Bioremediation

1 Center for Genomic Gastronomy, *Smog Tasting* (artist book, 2012).

2 Ellen Swallow Richards was the first woman to be granted a degree from and subsequently teach at the Massachusetts Institute of Technology. She taught 'sanitary chemistry' after her unpaid work in the Women's Laboratory ended with the laboratory's closure in 1883. We would recognise her laboratory as a public health laboratory today. See Robert Clarke, *Ellen Swallow: The Woman Who Founded Ecology* (Chicago: Follet, 1973). In their book *For Her Own Good: Two Centuries of the Experts' Advice to Women*, Barbara Ehrenreich and Deirdre English are careful to contextualise Richards's entry into Home Economics as 'not so much a triumph as a concession'; this new field was her only choice given her limited options as a woman in science in the late nineteenth century. Barbara Ehrenreich and Deirdre English, *For Her Own Good: Two Centuries of the Experts' Advice to Women* (New York: Knopf, 2013), p. 166.

3 Sanitation was influenced by eugenics and concepts of racial purity that animated many discourses of the late nineteenth-century United States (an era that saw tremendous numbers of immigrants arrive in the US). Although she unambiguously supported eugenics, albeit with the caveat that more scientific research must be done before implementing eugenic programmes (a position shared by many of her contemporaries), Richards distinguished between environmental remediation and the larger eugenics project. A set of practices she called 'euthenics' could be put in place immediately, with the goal of manipulating environmental rather than genetic factors to improve human lives. While her investment in eugenics cannot be ignored, her emphasis on ecological rather than genetic sanitation allows for connections between Richards's work and environmental art-science collaborations of the twentieth and twenty-first centuries. For a careful reading of Richards's investment in eugenics and racial purity, see Kristen Egan, 'Conservation and cleanliness: racial and environmental purity in Ellen Richards and Charlotte Perkins Gilman', *Women's Studies Quarterly* 39/3–4 (2011), pp. 77–92.

4 For example, Nicholas Mirzoeff's reading of Monet's *Impression: Sun Rising* (1873) makes the compelling argument that 'this is a painting that at once reveals and aestheticizes anthropogenic environmental destruction'. Nicholas Mirzoeff, 'Visualizing the anthropocene', *Public Culture* 26/2 (2014), pp. 213–32.

5 Robert Mitchell, *Bioart and the Vitality of Media* (Seattle, WA: Washington University Press, 2010).

6 Helen Mayer Harrison and Newton Harrison, *Shrimp Farm, Survival Piece #2*. Available at http://goo.gl/2xzuHw (accessed 10 December 2014). See also Leslie Ryan, 'Art + ecology: land reclamation works of artists Robert Smithson, Robert Morris, and Helen Mayer Harrison and Newton Harrison', *Environmental Philosophy* 4/1 and 2 (2007), pp. 95–116 and Craig Adcock, 'Conversational drift: Helen Mayer Harrison and Newton Harrison', *Art Journal* 51/2 (1992), pp. 35–45.

7 Adcock, 'Conversational drift', p. 35.

8 Ryan, 'Art + ecology', p. 106.

9 Maurice Tuchman, *A Report on the Art and Technology Program of the Los Angeles County Museum of Art* (Los Angeles, CA: LACMA, 1971). Available at https://archive.org/stream/reportonarttechn00losa_/reportonarttechn00losa#page/n0/mode/2up (accessed 22 January, 2015).

10 Harrisons, *Shrimp Farm, Survival Piece #2*.

11 Adcock, 'Conversational drift', p. 43.

12 Ellen Richards, 'Report of the Massachusetts Board of World's Fair Managers' (Boston, 1983). Available at http://libraries.mit.edu/archives/exhibits/esr/esr-rumford.html (accessed 19 January 2015).

13 Ryan, 'Art + ecology', p. 97.

14 Adcock, 'Conversational drift', p. 37.

15 Among many exhibitions and books dedicated to contemporary artists working with ecological systems, see Alan Sonfist (ed.), *Art in the Land: A Critical Anthology of Environmental Art* (New York: Plume, 1983); Grant Kester, *Groundworks: Environmental Collaboration in Contemporary Art* (Pittsburgh, PA: Carnagie Mellon University Press, 2005); Lucy Lippard, *Weather Report: Art and Climate Change* (Boulder, CO: Boulder Museum of Contemporary Art, 2007); Stephanie Smith and Victor Margolin (eds), *Beyond Green: Towards a Sustainable Art* (Chicago, IL: Smart Museum of Art, 2009); Linda Weintraub, *To Life!: Eco Art in Pursuit of a Sustainable Planet* (Berkeley, CA: University of California Press, 2012).

16 Tom Finkelpearl, 'Mel Chin on Revival Field', *Dialogues in Public Art* (Cambridge, MA: MIT Press, 2001), p. 385.

17 Adcock, 'Conversational drift', p. 43.

18 Finkelpearl, *Dialogues in Public Art*, p. 390.

19 Mel Chin with Rufus Chaney, *Revival Field* (1991–ongoing). Available at http://melchin.org/oeuvre/revival-field (accessed 1 December 2014). See also Zoë Ryan and Mel Chin, 'A conversation with Mel Chin', *Log* 8 (Summer 2006), pp. 59–68.

20 Tom Finkelpearl, 'Dr Rufus Chaney on Revival Field', *Dialogues in Public Art*, p. 411.

21 Donna Haraway, 'A Cyborg Manifesto: science, technology, and socialist-feminism in the late twentieth century', in her *Simians, Cyborgs, and Women* (New York: Routledge, 1991), p. 176.
22 Adcock, 'Conversational drift', p. 36.
23 Michael Marder, *Plant-Thinking: Toward a Philosophy of Vegetal Life* (New York: Columbia, 2013), pp. 8–9.
24 Nicola Twilley, 'Cross-species dining: an interview with Natalie Jeremijenko and Mihir Desai', *Edible Geography*, October 27, 2010. Available at http://www.ediblegeography.com/cross-species-dining-an-interview-with-natalie-jeremijenko-and-mihir-desai/ (accessed 13 February 2015).
25 Twilley, 'Cross-species dining'.
26 Natalie Jeremijenko, *X-Species: The Adventure Club* brochure. Available at http://www.environmentalhealthclinic.net/xooz/wp-content/uploads/2010/08/100721_xspecies-adventureclub.pdf (accessed 13 February 2015).
27 United States Environmental Protection Agency, 'How people are exposed to mercury'. Available at http://www.epa.gov/hg/exposure.htm (accessed 23 May 2014).
28 Chitinase is an effective chelating agent because it is high in nitrogen. For more information on different kinds of chitinase, see Hamid et al., 'Chitinases: an update', *Journal of Pharmacy and Bioallied Sciences* 5/1 (January–March 2013), pp. 21–9.
29 Twilley, 'Cross-species dining'.
30 As an indicator of its populist potential, a 2008 *Make Magazine* molecular gastronomy tutorial featured reverse spherification. See Michael Zbyszynski, 'Molecular gastronomy: spherify your food for a new culinary experience', *Make Magazine* 14 (2008), pp. 149–52.
31 Twilley, 'Cross-species dining'.
32 Lindsay Kelley, 'Digesting wetlands: cooking and eating across species with Natalie Jeremijenko's *Cross(x)Species Adventure Club*', forthcoming in Silvia Bottinelli and Margherita d'Ayala (eds), *What's Cooking? Food, Art and Counterculture* (Fayetteville, AR: University of Arkansas Press, 2015).
33 Donna Haraway, 'Multispecies cosmopolitics: staying with the trouble', 2013 Distinguished Lecture, Arizona State University, 5 March 2013.
34 Isabel Stengers, *Cosmopolitics II*, trans. Robert Bononno (Minneapolis, MN: Minnesota University Press, 2011), pp. 371–2.
35 Ellen Swallow Richards, quoted in Pamela C. Swallow, *The Remarkable Life and Career of Ellen Swallow Richards: Pioneer in Science and Technology* (New York: Wiley, 2014), pp. 89–90.
36 Ibid., p. 89.
37 Massachusetts Board of Health, *Massachusetts Board of Health Guidebook*, pp. 10–2. Available at http://www.mahb.org/Library/Guide%20book/gbook10.pdf (accessed 13 February 2015).

38 See Katherine Harmon, *The Map as Art: Contemporary Artists Explore Cartography* (Princeton, NJ: Princeton Architectural Press, 2010) and Karen O'Rourke, *Walking and Mapping: Artists as Cartographers* (Cambridge, MA: MIT Press, 2013).
39 Center for Genomic Gastronomy, 'LOCI food lab: PDX summary'. Available at http://genomicgastronomy.com/blog/loci-food-lab-pdx-summary/ (accessed 21 January 2015).
40 Adcock, 'Conversational drift', p. 37.

11 Plumpiñon

1 Stengers precedes this question with other useful questions that animate my interest in how food, art, ethnobotany and conceptions of humanitarian aid coincide: 'How can protagonists capable of complicating the problem be empowered? How can the presence of those who might share in the associated risks, choices, and decisions be ensured?' Isabel Stengers, *Cosmopolitics II*, trans. Robert Bononno (Minneapolis, MN: Minnesota University Press, 2011), p. 368.
2 This recipe and accompanying essay about *Plumpiñon* was first developed as part of my MFA thesis at the University of California Santa Cruz and then was refined for publication in Eben Kirksey (ed.), *The Multispecies Salon* (Chapel Hill, NC: Duke University Press, 2014), pp. 122–134.
3 Michael Wines, 'Hope for hungry children, arriving in a foil packet', *New York Times*, 8 August 2005. Available at http://www.nytimes.com/2005/08/08/international/africa/08niger.html (accessed 25 September 2014).
4 For example: 'Hungry people have started adding "bitter" berries to their diet – this is survival food, normally unpalatable but when starving, the unpalatable becomes welcome – essential.' Alastair Stewart, 'Millions face starvation as Niger prays in vain for rain', *The Independent*, 9 July 2010. Available at http://www.independent.co.uk/news/world/africa/millions-face-starvation-as-niger-prays-in-vain-for-rain-2022190.html (accessed 20 January 2015); 'Some of the people have been surviving on the bitter berries from the local Anza bush, which are usually considered inedible.' Emma Hurd, 'UN pleads for £124m to help starving Niger', *Sky News*, 30 April 2010. Available at http://news.sky.com/story/775823/un-pleads-for-124m-to-help-starving-niger (accessed 20 January 2015); 'Facing the worst harvest in years and surviving on the bitter berries and tasteless grasses that grow in the surrounding sandy soil.' Robyn Dixon, 'Season of destruction returns to Niger', *Los Angeles Times*. Available at http://articles.latimes.com/2005/aug/07/world/fg-niger7 (accessed 20 January 2015). Not everyone dismisses the Anza bush: *The Lost Crops of Africa*, a National Research Council publication devoted to 'evaluating underexploited African

plant resources that could help broaden and secure Africa's food supply', has a more positive spin on bitter berries: 'Having survived thousands of years of recurrent drought without horticultural help this wild species holds the potential to make life more bearable under the desiccating conditions in which millions of Africa's most destitute are increasingly forced to exist [...] The species produces enough different products to sustain human life almost by itself.' National Research Council, *The Lost Crops of Africa, Volume III: Fruits* (Washington DC: National Academies Press, 2008), p. 221.

5 See Justin's nut butters. Available at http://www.justinsnutbutter.com/ (accessed 23 January 2014) for details on a range of products that include chocolate peanut butter, honey peanut butter and maple almond butter all distributed in foil packets.

6 Anza berries are often presented with a warning, as in the text accompanying Jane Hahn's photograph of Nigerien women: 'Anza berries [...] can be highly toxic if not prepared carefully.' The group of women we see in the photograph, one of whom displays a handful of Anza berries, is implicated in this statement. Will the woman holding these berries 'not prepare carefully'? The berries become a dangerous, risky food, a non-food, and the scarce handful implies that everyone in the photo will have to share this quantity of berries for lunch. A reductive nutritional reading of the Anza berry alongside the familiar, friendly peanut would be inappropriate, given the berry's status as fringe food and the peanut's established entry into global food ways. Organisations like the United States National Research Council and the Belgian Royal Museum for Central Africa act as documentarians, often gleaning information about the Anza berry from veterinary contexts. Jane Hahn, 'Niger – food crisis'. Available at http://www.janehahn.com/#/nigers-food-crisis (accessed 4 May 2015).

7 Anthropologist Tom D. Dillehay recently determined that the domesticated peanut has a more established presence in the Andes than previously thought. His team found evidence of agricultural peanut activities 10,000 years ago, even though 'the peanut was long thought to be among the later cultivated plants of the Andes, and one that is particularly suited to the lowland tropical forests and savannahs where it was prized as a high-protein complement to starchy manioc-based diets'. Dillehay locates a geographic centre for the peanut: 'The peanut's center of origin is believed to be in an area east of the Andes comprising southeastern Bolivia, north-western Argentina, northern Paraguay, and the western Mato Grosso region of Brazil.' Tom Dillehay, Jack Rossen, Thomas C. Andres and David E. Williams, 'Preceramic adoption of peanut, squash, and cotton in Northern Peru', *Science* **316** (2007), pp. 1890–3, p. 1890. The peanut has messy origins, with this initial 'centre' having been productively polluted by other trade ways. In popular cooking culture in North and South America, the peanut is often promoted as a food that may 'help recover the health of Indigenous peoples'. American

Indian Health and Diet Project, 'Foods indigenous to the Western hemisphere', n.d. Available at http://www.aihd.ku.edu/foods/peanuts.html (accessed 25 January 2015).

8 Bruce Kershner and Craig Tufts, *National Wildlife Federation Field Guide to Trees of North America* (New York: Sterling, 2008), p. 92.

9 Pinyon is an Anglicization of 'piñon' that stops short of 'pine'. 'Pinyon' is an effort to write the word without the 'ñ' while preserving its Spanish pronunciation.

10 There are many natural histories of the piñon tree. For three good examples among many, see Lisa Floyd, *Ancient Piñon-Juniper Woodlands: A Natural History of Mesa Verde Country* (Boulder, CO: University of Colorado Press, 2003), Alfred Savinelli, *Plants of Power: Native American Ceremony and the Use of Sacred Plants* (Summertown, TN: Native Voices, 2002) and David Rhode, *Native Plants of Southern Nevada: An Ethnobotany* (Salt Lake City, UT: University of Utah Press, 2002). One of my favourites is Ronald Lanner's *The Piñon Pine: A Natural and Cultural History* (Reno, NV: University of Nevada Press, 1981). Lanner's text includes a 'section on pine nut cookery' authored by Harriette Lanner. The book joins a range of texts that Lanner has produced in an effort to educate people about trees. The cookery section illuminates the rich interconnections between kitchens, forests and human animals.

11 Heather Paxson, 'Post-Pasteurian cultures: the microbiopolitics of raw-milk cheese in the United States', *Cultural Anthropology* 23/1 (2008), pp. 15–47, p. 16 (emphasis in original).

12 Ibid., p. 18.

13 For documentation of a hot-press process designed to manufacture fused plastic at a bigger scale than my own efforts, see Waste for Life. website Available at http://wasteforlife.org/ (accessed 25 January 2015). For examples of fused plastic design prototypes from the Rhode Island School of Design's 'It's in the Bag' design competition see RISD Waste for Life website. Available at http://risdwasteforlife.wordpress.com/ (accessed 25 January 2015).

14 Birdie Jaworski, 'Pride in piñon', in *My Tiny Vegas: Life between the Sangre de Christo Mountains and the Great Plains* (Create Space, 2009).

15 Vernon Mayes and Barbara Bayless Lacy, *Nanise': A Navajo Herbal* (Tsaile, AZ: Navajo Community College Press, 1989).

16 The Navajo Ethnobotany project, produced by Dykeman Roebuck Archaeology, has collected data from Maye and Bayless and other sources to produce a study detailing significant native plants found in Navajo lands. See Dykeman Roebuck Archaeology, 'Piñon pine'. Available at http://www.drarchaeology.com/herbal/ethno/pinyon.htm (accessed 25 January 2015) as well as Mayes and Bayless Lacy *Nanise'*.

17 Rebecca McLain, 'Management guidelines for expanding pinyon nut production in Colorado's pinyon-juniper woodlands', Institute for Culture and Ecology, p. 7.

18 'Navajos often harvest as families, and they have a reputation for being the most skilled pinyon nut pickers. One buyer in northern New Mexico said, "Most commercial pickers will pick five pounds where a Navajo will pick 30 pounds. They've got a system in place. Know-how is a big part of doing well in the business." [...] For the Navajo, Western Shoshone, and Paiute – and likely other Native American pickers – the pinyon harvest is a sacred activity, even when the nuts are harvested for commercial exchange. Most Navajos, for example, give an offering and say a blessing as part of the harvesting ritual.' Ibid., p. 38.

19 Jaworski, 'Pride in piñon', p. 12.

20 Madrid has a fascinating, if brief, history, with Superintendent of Mines Oscar Huber providing high living standards for town workers and compulsory co-operative efforts around recreational activities including annual holiday light displays. Huber's son Joe started renting abandoned mining properties to local artists in the 1970s. The most extensive history of Madrid I have found online is Kathy Weiser, 'New Mexico legends: Madrid – a ghost town reborn', *Legends of America*, September 2011. Available at http://www.legendsofamerica.com/nm-madrid.html (accessed 29 January 2015). Also see Kathryn Hovey's history of the town's reinvention, *Anarchy and Community in the New American West* (Albuquerque, NM: University of New Mexico Press, 2005).

21 Mexicans were counted as white in the US census until 1930. The statistic I use here comes from the 2010 census, where 'Hispanic or Latino' can be selected alongside race and 'type', which refers to national origin. United States Census Bureau, American Fact Finder, 'Hispanic or Latino by Type: 2010 Census Summary File 1', *American Fact Finder*. Available at http://factfinder2.census.gov/faces/tableservices/jsf/pages/productview.xhtml?pid=DEC_10_SF1_QTP10&prodType=table (accessed 30 January 2015).

22 After posting my grandfather's eulogy online, my extended Narvaez family contacted me. We have since been corresponding about family history and genealogy. The name Narvaez appears to have died out due to a preponderance of women in our family. Upon learning that my branch of the family also neglected to produce a male heir to the name Kelley, my cousin Juanita Gordon wrote: 'It seems the Narvaez name may have been meant to die out.' My great-great-grandfather, Max Narvaez, appears in a list of names of miners, published by Midori Snyder as 'Madrid's Illiad: the list of miners' names: Part III.' Available at http://msnyder.typepad.com/the_labyrinth/2011/01/madrids-illiad-the-list-of-names.html (accessed 30 January 2015). My grandfather's eulogy is available at http://starvationseeds.blogspot.com/2010/01/piling-up.html (accessed 30 January 2015).

23 Lanner, *The Piñon Pine*, p. 89.

24 Donald Olson et al., 'Piñon pines and the route of Cabeza de Vaca', *Southwest Historical Quarterly* 101 (1997), pp. 174–86, p. 183.

25 'The term "cosmopolitics" introduces what is neither an activity, nor a negotiation, nor a practice, but the mode in which the problematic copresence of practices may be actualized: the experience, always in the present, of the one into whom the other's dreams, hopes, and fears pass.' Stengers, *Cosmopolitics II*, p. 372. Looking back on de Vaca, as readers of the *Journey* and its biogeography have noted, de Vaca appears to have described this mode in his encounters with the piñon.

26 Ibid., p. 372.

27 If the nuts de Vaca describes are *Pinus remota*, this makes a strong case for a southerly route, given *P. remota*'s biogeography. Biogeography also helps interpret the way the de Vaca party consumed the piñon seeds – crushing *P. edulis* seeds into a powder would be difficult because their shells do not break down. Cabeza de Vaca, *The Narrative of Cabeza de Vaca*, trans. Rolena Adorno and Patrick Charles Pautz (Lincoln, NE: University of Nebraska Press, 2003), p. 140. The mashed green piñon paste reminds me of the delivery system used with both Plumpy'nut and 'Plumpiñon'.

28 Mary Louise Pratt, *Imperial Eyes: Travel Writing and Transculturation* (New York: Routledge, 2008), p. 78.

29 James Clifford, *Routes: Travel and Translation in the Late Twentieth Century* (Cambridge, MA: Harvard University Press, 1997), p. 194.

30 Pratt, *Imperial Eyes*, pp. 85, 81.

31 de Vaca, *Narrative*, p. 141.

32 Stengers, *Cosmopolitics II*, p. 372.

33 Lanner, *The Piñon Pine*, p. 97.

Epilogue: Dysphagiac

1 Lucy Duperon, 'Lucy's real food for the tube' (2008). Available at https://web.archive.org/web/20110517011133/http://lucysrealfood.com/vegetables.htm (accessed 19 March 2014).

2 Amelie Hastie, reflecting on her life since she lost her sense of smell, writes that, 'I dreamed [...] of writing a memoir called "Occasional Essays of an Anosmiac". The proper term for a person who cannot smell is actually "anosmic", but I prefer my version, as it offers a maniacal quality to my new identity'. With Hastie, I prefer dysphagiac for the same reasons. Responses to an inability to swallow or eat are inevitably maniacal: the constant need to spit phlegm into a cup because it cannot be swallowed, the unconscious reach for and swallow of popcorn, the tube's relentless, careful appetite. These are each small manias. Amelie Hastie, 'Senseless eating', *parallax* 66 (2013), pp. 64–73, p. 64.

3 Michael Pollan, *In Defense of Food: An Eater's Manifesto* (New York: Penguin, 2008), p. 157.

4 Lucy Duperon, 'Making your own food for tube feeding, homemade food for G-tube'. Available at http://lucysrealfood.com/ (accessed 15 November 2012) (link no longer working). Duperon has since taken her site offline, having determined that there were sufficient resources available for caretakers interested in homemade tube food.

5 Linda Lee, 'Bridal hunger games: losing weight in time for the wedding', *New York Times*, 13 April 2012. Available at http://www.nytimes.com/2012/04/15/fashion/weddings/Losing-Weight-in-Time-for-the-Wedding.html?_r=0 (accessed 19 February 2015).

6 Margaret Atwood, *Oryx and Crake* (New York: Anchor, 2004), p. 158.

7 Herbert George Wells, *The Sleeper Awakes* (Lincoln, NE: Nebraska University Press 2000), p. 26.

Bibliography

Adams, William Mark and Martin Mulligan (eds), *Decolonizing Nature: Strategies for Conservation in a Post-colonial Era* (London: Earthscan, 2012).

Adcock, Craig, 'Conversational drift: Helen Mayer Harrison and Newton Harrison', *Art Journal* 51/2 (1992), pp. 35–45.

Alpers, Svetlana, 'The studio, the laboratory, and the vexations of art', in Caroline Jones and Peter Galison (eds), *Picturing Science Producing Art* (New York: Routledge, 1998), pp. 401–17.

American Indian Health and Diet Project, 'Foods indigenous to the Western hemisphere', n.d. Available at http://www.aihd.ku.edu/foods/peanuts.html (accessed 25 January 2015).

Anker, Suzanne, *The Molecular Gaze: Art in the Genetic Age* (Cold Spring Harbor, NY: Cold Spring Harbor Laboratory Press, 2004).

Annas, George J., 'Bioterror and "bioart" – a plague o' both your houses', *The New England Journal of Medicine* 354/25 (2006), pp. 2715–20.

Arnold, Ron, *Ecoterror: The Violent Agenda to Save Nature: The World of the Unabomber* (Bellevue, WA: Free Enterprise Press, 1997).

Ashton, Kevin, 'That "internet of things" thing', *RFiD Journal*, 22 June 2009. Available at http://www.rfidjournal.com/articles/view?4986 (accessed 23 January 2014).

Attar, Dena, *Wasting Girls' Time: The History and Politics of Home Economics* (London: Virago, 1990).

Atwood, Margaret, *Oryx and Crake* (New York: Doubleday, 2003).

Aurora, Valerie, 'Valerie Aurora meets the Honeywell H316 Kitchen Computer', n.d. Available at http://valerieaurora.org/kitchen.html (accessed 4 February 2015).

Avakian, Arlene Voski and Barbara Haber (eds), *From Betty Crocker to Feminist Food Studies: Critical Perspectives on Women and Food* (Boston, MA: Massachusetts University Press, 2005).

Baille Gerritsen, Vivienne, 'The greenest of us all?', *Protein Spotlight* 11 (June 2001). Available at http://web.expasy.org/spotlight/pdf/sptlt011.pdf (accessed 15 September 2012).

Baker, Steve (ed.), *Picturing the Beast: Animals, Identity, and Representation* (Champaign, IL: Illinois University Press, 2001).

—'Sloughing the human', in Cary Wolfe (ed), *Zoontologies* (Minneapolis, MN: Minnesota University Press, 2003), pp. 147–64.

—*Artist / Animal* (Minneapolis, MN: Minnesota University Press, 2003).

—and Carol Gigliotti, 'We have always been transgenic', *AI and Society* 20 (2006), pp. 35–48.

Bakke, Monica, 'Zoe-philic desires: wet media art and beyond', *parallax* 14/3 (2008), pp. 21–34.

Barthes, Roland, *Mythologies* (New York: Hill and Wang, 1972).

—*Image, Music, Text*, trans. Stephen Heath (New York: Hill and Wang, 1977).

Bartkowski, Francis, *Kissing Cousins: A New Kinship Bestiary* (New York: Columbia University Press, 2008).

Beecher, Catherine, *A Treatise on Domestic Economy: For the Use of Young Ladies at Home and at School* (Boston, MA: Thomas Webb, 1843).

Belasco, Warren, *Meals to Come: A History of the Future of Food* (Berkeley, CA: California University Press, 2006).

Bell, Genevieve, Mark Blythe and Phoebe Sengers, 'Making by making strange: defamiliarization and the design of domestic technologies', *ACM Transactions on Computer-Human Interaction* 12/2 (2005), pp. 149–73.

Bennett, Jane, 'The agency of assemblages', *Vibrant Matter: A Political Ecology of Things* (Durham, NC: Duke University Press, 2010).

Bergson, Henri, *Matter and Memory* (New York: Zone, 1991).

Berlage, Nancy, 'The establishment of an applied social science: Home Economists, Science, and Reform at Cornell University, 1870–1930', in Helen Silverberg (ed.), *Gender and American Social Science: The Formative Years* (Princeton, NJ: Princeton University Press, 1998).

Birringer, Johannes, 'Performance and science', *PAJ: A Journal of Performance and Art* 29 1 (2007), pp. 21–35.

Bishop, Claire (ed.), *Participation: Documents in Contemporary Art* (London/Cambridge: Whitechapel Gallery/MIT Press, 2006).

Boltanski, Christian, Bertrand Lavier and Hans Ulrich Obrist, 'DO-IT at e-flux', e-flux.com. Available at http://www.e-flux.com/projects/do_it/homepage/do_it_home.html (accessed 30 January 2014).

Bower, Anne (ed.), *Recipes for Reading: Community Cookbooks, Stories, Histories* (Amherst, MA: Massachusetts University Press, 1997).

Broadhurst, Susan and Josephine Machon (eds), *Performance and Technology: Practices of Virtual Embodiment and Interactivity* (New York: Palgrave, 2006).

Brodwin, Paul (ed.), *Biotechnology and Culture: Bodies, Anxieties, Ethics* (Bloomington, IL: Indiana University Press, 2000).

Brooklyn Museum, 'Brooklyn Museum: Elizabeth A. Sackler Center for Feminist Art: *The Dinner Party*', n.d. Available at http://www.brooklynmuseum.org/eascfa/dinner_party/ (accessed 10 February 2015).

Broude, Norma and Mary D. Garrard (eds), *The Power of Feminist Art* (New York: Harry N. Abrams, 1994).

Bryan-Wilson, Julia, 'The political problem of luck', *Plazm* (2006), pp. 22–30.

Bulatov, Dimitry (ed.), *Biomediale: Contemporary Society and Genomic Culture*. (Kaliningrad: Yantarny Skaz, 2004).

Bunt, Stuart, 'The role of the scientist and science in bio-art', in Melentie Pandilovski (ed.), *Art in the Biotech Era* (Adelaide: Experimental Art Foundation, 2008), pp. 62–7.

Callicott, J. Baird and Michael P. Nelson (eds), *The Great New Wilderness Debate* (Athens, GA: Georgia University Press, 1998).

Carrel, Alexis, 'Physiological time', *Science* 74/1929 (1931), pp. 618–21.

—and Charles Lindbergh, *The Culture of Organs* (New York: Hoeber, 1938).

Carruth, Allison, 'Culturing Food: Bioart and In Vitro Meat', *parallax* 19/1 (2013), pp. 88–100.

Carson, Rachel, *Silent Spring* (New York: Houghton Mifflin, 2002).

Catts, Oron, 'Biotech art workshop conducted by Symbiotica', presented at the conference *Bio-Art and the Public Sphere*, University of California, Irvine, 10–14 October 2005.

—'Semi living art', in Eduardo Kac (ed.), *Signs of Life* (Cambridge, MA: MIT Press, 2007).

—'Labs shut open: a biotech hands-on workshop for artists', in Beatriz da Costa and Kavita Philips (eds), *Tactical Biopolitics Art, Activism, and Technoscience* (Cambridge, MA: MIT Press, 2008), pp. 143–56.

—'Disembodied livestock: the promise of a semi-living utopia', *parallax* 19/1 (2013), pp. 101–13.

—'Towards a new class of being: the extended body', *Artnodes* 6 (2006). Available at http://www.uoc.edu/artnodes/6/dt/eng/catts_zurr.pdf (accessed 7 May 2015).

—and Ionat Zurr, 'Growing semi-living sculptures: The Tissue Culture & Art Project', *Leonardo* 35/4 (2002), pp. 365–70.

Center for Genomic Gastronomy, 'Glowing sushi', n.d. Available at http://www.glowingsushi.com/ (accessed 15 September 2012).

Center for Genomic Gastronomy, 'LOCI food lab: PDX summary'. Available at http://genomicgastronomy.com/blog/loci-food-lab-pdx-summary/ (accessed 21 January 2015).

Center for Genomic Gastronomy, *Smog Tasting* (artist book, 2012).

Centre for Critical and Cultural Theory, Cardiff University, 'Zoontotechnics (animality/technicity)'. Available at http://www.cardiff.ac.uk/encap/newsand-events/events/conferences/zoontotechnics.html (accessed 7 February 2014).

Chicago, Judy, *The Dinner Party: A Symbol of Our Heritage* (Garden City, NY: Anchor Books, 1979).

Chin, Christine, *The Genetically-Modified Foods Cook Book* (2004). Available at http://gmcookbook.com/ (accessed 18 March 2014).

—*Vegetable Human Hybrids*, n.d. Available at http://christinechin.net/hybrids/index.html (accessed 18 March 2014).

Chin, Mel with Rufus Chaney, *Revival Field* (1991–ongoing). Available at http://melchin.org/oeuvre/revival-field (accessed 1 December 2014).

Churchill, Winston, 'Fifty years hence', *Popular Mechanics*, March 1932, pp. 390–7.

Clarke, Robert, *Ellen Swallow: The Woman Who Founded Ecology* (Chicago, IL: Follet, 1973).

Clifford, James, *Routes: Travel and Translation in the Late Twentieth Century* (Cambridge, MA: Harvard University Press, 1997).

Compassion Over Killing, '45 days in hell: the life and death of a broiler chicken'. Available at http://www.chickenindustry.com/cfi/documentary/ (accessed 10 March 2014).

Computer History Museum, 'Bytes for bites: the kitchen computer', n.d. Available at http://www.computerhistory.org/revolution/minicomputers/11/362 (accessed 23 January 2014).

Cook, Lynne and Peter Wollen (eds), *Visual Display: Culture Beyond Appearances* (Seattle, WA: Bay Press, 1995).

Cooper, Melinda, Andrew Goffey and Anna Munster, 'Biopolitics, for now', *Culture Machine* 7 (2005). Available at http://www.culturemachine.net/index.php/cm/article/view/24/31 (accessed 22 April 2015).

Costello, Elizabeth K., Les Dethlefsen, David A. Relman, Keaton Stagaman and Brendan Bohannan, 'The application of ecological theory toward an understanding of the human microbiome', *Science* 336/6086 (2012), pp. 1255–62.

Cottiers-Pacella, Wendy, *The Bare Truth: Living GMO-Free in the 21st Century* (Fort Lauderdale, FL: Positive Nutrition, 2014).

Cowan, Ruth Schwartz, *More Work for Mother: The Ironies of Household Technology from the Open Hearth to the Microwave* (New York: Basic Books, 1983).

Crary, Jonathan and Sanford Kwinter (eds), *Incorporations* (New York: Zone, 1992).

Critical Art Ensemble, *Electronic Civil Disobedience and Other Unpopular Ideas* (Brooklyn, NY: Autonomedia, 1996).

—Jon McKenzie and Rebecca Schneider, 'Critical Art Ensemble: tactical media practitioners', *The Drama Review* 44/4 (2000), pp. 136–50.

—'Performing a cult', *The Drama Review* 44/4 (2000), pp. 167–73.

—'Recombinant theatre and digital resistance', *The Drama Review* 44/4 (2000), pp. 151–66.

—*Digital Resistance: Explorations in Tactical Media* (Brooklyn, NY: Autonomedia, 2001).

—*Molecular Invasion* (Brooklyn, NY: Autonomedia, 2002).

—'Fuzzy biological sabotage', 2013. Available at http://critical-art.net/MolecularInvasion.html (accessed 19 February 2015).

—Beatriz da Costa and Shyh-shiun Shyu, *Free Range Grain*, n.d. Available at http://critical-art.net/Original/free/freeRange.swf (accessed 4 February 2014).

Crocker, Betty, *Betty Crocker's Picture Cookbook* (Minneapolis, MN: Wiley and General Mills, 1950).

da Costa, Beatriz, 'The art of biologies and environmentalism', University of California Santa Cruz Digital Art and New Media Festival symposium, 5 May 2006.

—'Reaching the limit: when art becomes science', in Beatriz da Costa and Kavita Philips (eds), *Tactical Biopolitics* (Cambridge, MA: MIT Press, 2008), pp. 365–82.

—*Anti Cancer Survival Kit*. Available at http://www.rockethub.com/projects/12551-the-anti-cancer-survival-kit#description-tab (accessed 30 January 2014).

—and Claire Pentecost, 'Of Patriots and profits – new tools for keeping academic research in line', *Radical History Review* 93 (2005), pp. 111–21.

—and Kavita Philip (eds), *Tactical Biopolitics: Art, Activism, and Technoscience* (Cambridge, MA: MIT Press, 2008).

DaSilva, Edgar, 'Art, biotechnology and the culture of peace', *Electronic Journal of Biotechnology* 7/2 (2004), pp. 130–66.

Datar, Isha and Robert Bolton, 'The carnery', in Next Nature Network, *In Vitro Meat Cookbook* (Amsterdam: BIS Publishers, 2014), pp. 149–159.

Department of Human Development and Family Studies, Cornell University, 'Why the change to human ecology?' (2001). Available at http://rmc.library.cornell.edu/homeEc/cases/namechange.html (accessed 14 April 2012).

Derrida, Jacques, *Writing and Difference*, trans. Alan Bass (Chicago, IL: Chicago University Press, 1980).

Derrida, Jacques with Jean-Luc Nancy, '"Eating well", or the calculations of the subject: an interview with Jacques Derrida', in Eduardo Cadava, Peter Connor and Jean-Luc Nancy (eds), *Who Comes after the Subject?* (New York: Routledge, 1991).

de Vaca, Cabeza, *The Narrative of Cabeza de Vaca*, trans. Rolena Adorno and Patrick Charles Pautz (Lincoln, NE: University of Nebraska Press, 2003).

Dewey, John, *The Public and Its Problems* (Athens, OH: Swallow, 1991).

Didur, Jill, 'Re-embodying technoscientific fantasies: posthumanism, genetically modified foods, and the colonization of Life', *Cultural Critique* 53 (Winter 2003), pp. 98–115.

Dillehay, Tom, Jack Rossen, Thomas C. Andres and David E. Williams, 'Preceramic adoption of peanut, squash, and cotton in Northern Peru', *Science* 316 (2007), pp. 1890–3.

Dixon, Robyn, 'Season of destruction returns to Niger', *Los Angeles Times*. Available at http://articles.latimes.com/2005/aug/07/world/fg-niger7 (accessed 20 January 2015).

Dobson, Andrew (ed.), *Fairness and Futurity: Essays on Environmental Sustainability and Social Justice* (Oxford: Oxford University Press, 1999).

Doing, Park, '"Lab hands" and the "scarlet O": epistemic politics and (scientific) labor', *Social Studies of Science* 34 (2004), pp. 299–323.

—'Give me a laboratory and I will raise a discipline: the past, present, and future politics of laboratory studies in STS', in Edward Hackett, Olga Amsterdamska, Michael Lynch and Judy Wajcman (eds), *The Handbook of Science and Technology Studies*, 3rd edn (Cambridge, MA: MIT Press, 2008), pp. 279–98.

Dougherty, Dale, 'Maker: Natalie Jeremijenko', *Make Magazine* 2 (2005).

Duperon, Lucy, 'Lucy's real food for the tube' (2008). Available at https://web.archive.org/web/20110517011133/http://lucysrealfood.com/vegetables.htm (accessed 19 March 2014).

—'Making your own food for tube feeding, homemade food for G-tube'. Available at http://lucysrealfood.com/ (accessed 15 November 2012)(link no longer working).

Durant, Elizabeth, 'A lab of their own', *MIT Technology Review*. Available at http://libraries.mit.edu/archives/exhibits/esr/esr-womenslab.html (accessed 9 March 2014).

Dykeman Roebuck Archaeology, 'Piñon pine'. Available at http://www.drarchaeology.com/herbal/ethno/pinyon.htm (accessed 25 January 2015).

Eamons, William, *Science and the Secrets of Nature: Books of Secrets in Medieval and Early Modern Culture* (Princeton, NJ: Princeton University Press, 1994).

Edelman, P.D., D.C. McFarland, V.A. Mironov and J.G. Matheny, '*In vitro*-cultured meat production', *Tissue Engineering* 11/5–6 (2005), pp. 659–62.

Egan, Kristen, 'Conservation and cleanliness: racial and environmental purity in Ellen Richards and Charlotte Perkins Gilman', *Women's Studies Quarterly* 39/3–4 (2011), pp. 77–92.

Ehrenreich, Barbara and Deirdre English, *For Her Own Good: Two Centuries of the Experts' Advice to Women* (New York: Anchor, 1978).

Elias, Megan, *Stir it Up: Home Economics in American Culture* (Philadelphia, PA: Pennsylvania University Press, 2010).

'Ellen Swallow Richards and MIT'. Available at http://libraries.mit.edu/archives/exhibits/esr/esr-mit.html (accessed 9 March 2014).

Epstein, Barbara, *The Politics of Domesticity* (Middletown, CT: Wesleyan University Press, 1981).

Ettlinger, Steve, *Twinkie, Deconstructed: My Journey to Discover How the Ingredients Found in Processed Foods Are Grown, Mined (Yes, Mined), and Manipulated into What America Eats* (New York: Penguin, 2007).

European Commission, 'Directive 2001/18/EC of the European Parliament on the deliberate release of genetically modified organisms', 12 March 2001. Available at http://eur-lex.europa.eu/LexUriServ/LexUriServ.do?uri=CELEX:32001L0018:EN:HTML (accessed 4 February 2014).

Felshin, Nina (ed.), *But is it Art? The Spirit of Art as Activism* (Seattle, WA: Bay Press, 1995).

Finkelpearl, Tom, *Dialogues in Public Art* (Cambridge, MA: MIT Press, 2001).

Fisher, Mary Frances Kennedy, *With Bold Knife and Fork* (New York: Counterpoint, 1969).

Floyd, Janet and Laurel Forster (eds), *The Recipe Reader* (Burlington, VT: Ashgate, 2003).

Floyd, Lisa, *Ancient Piñon-Juniper Woodlands: A Natural History of Mesa Verde Country* (Boulder, CO: University of Colorado Press, 2003).

Foley, Robert, 'Are we the only animals to cook our food?', 16 November 2008. Available at http://www.thenakedscientists.com/HTML/content/latest-questions/question/2169/ (accessed 15 September 2012).

'Food for thought (and for your face)'. Available at http://food4thoughtand4urface.wordpress.com/tag/oryx-and-crake/ (accessed 29 January 2014).

The Forme of Cury. Available at http://www.pbm.com/~lindahl/foc/ (accessed 7 February 2014).

Franklin, Sarah and Margaret Lock (eds), *Remaking Life and Death: Toward an Anthropology of the Biosciences* (Santa Fe, NM: School of American Research Press, 2003).

Frederick, Christine, *Household Engineering: Scientific Management in the Home* (Chicago, IL: American School of Home Economics, 1920).

—*Selling Mrs Consumer* (New York: The Business Bourse, 1929).

Freedberg, David, 'Iconography between the History of Art and the History of Science: art, science, and the case of the urban bee', in Caroline Jones and Peter Galison (eds), *Picturing Science Producing Art* (New York: Routledge, 1998), pp. 272–96.

'Fridge sends spam emails as attack hits smart gadgets', *BBC News*, 17 January 2014. Available at http://www.bbc.co.uk/news/technology-25780908 (accessed 23 January 2014).

Fried, Ian and Melanie Austria Farmer, '3Com lets Audrey out the door', *CNET*, 17 October 2000. Available at http://news.cnet.com/3Com-lets-Audrey-out-the-door/2100-1040_3-247152.html (accessed 23 January 2014).

Friedman, David, *The Immortalists: Charles Lindbergh, Dr Alexis Carrel, and Their Daring Quest to Live Forever* (New York: HarperCollins, 2007).

Friedman, Ken, Owen Smith and Lauren Sawchyn (eds), *Fluxus Performance Workbook*, *Performance Research* 7/3 (2002). Available at http://www.thing.net/~grist/ld/fluxusworkbook.pdf (accessed 18 March 2014).

Gessert, George, *Green Light: Toward an Art of Evolution* (Cambridge, MA: MIT Press, 2010).

Gigliotti, Carol, 'Leonardo's choice: the ethics of artists working with genetic technologies', *AI and Society* 20 (2006), pp. 22–34.

Gitlin, Jonathan M., 'Is this sausage supposed to be green?', 12 January 2006. Available at http://arstechnica.com/science/news/2006/01/2477.ars (accessed 15 September 2012).

Goldstein, Carolyn, 'Part of the package: Home Economists in consumer products industries, 1920–1940', in Sarah Stage and Virginia B. Vincenti (eds), *Rethinking Home Economics: Women and the History of a Profession* (Ithaca, NY: Cornell University Press, 1997), pp. 271–96.

Gómez-Peña, Guillermo, 'Disclaimer', *The Drama Review* 50/1 (Spring 2006), pp. 149–58.

Grant, Julia, 'Modernizing mothers: Home Economics and the parent education movement, 1920–1945', in Sarah Stage and Virginia B. Vincenti (eds), *Rethinking Home Economics: Women and the History of a Profession* (Ithaca, NY: Cornell University Press, 1997), pp. 55–76.

Grosz, Elizabeth, *Chaos, Territory, Art: Deleuze and the Framing of the Earth* (New York: Columbia University Press, 2008).

Gunn, James (ed.), *The Road to Science Fiction*, vol. 2: *From Wells to Heinlein* (Lanham, MD: Scarecrow, 2002).

Hackett, Edward, Olga Amsterdamska, Michael Lynch and Judy Wajcman (eds), *The Handbook of Science and Technology Studies*, 3rd edn (Cambridge, MA: MIT Press, 2008).

Hamid, Rifat, Minhaj A. Khan, Mahboob Ahmad, Saleem Javed, Malik Mobeen Ahmad, Malik Zainul Abdin, and Javed Musarrat, 'Chitinases: an update', *Journal of Pharmacy and Bioallied Sciences* 5/1 (January–March 2013), pp. 21–9.

Haraway, Donna, *Simians, Cyborgs, and Women* (New York: Routledge, 1991).

—*Modest_Witness@Second_Millennium.FemaleMan©_Meets_OncoMouse: Feminism and Technoscience* (New York: Routledge, 1997).

—*The Companion Species Manifesto: Dogs, People and Significant Otherness* (Chicago, IL: Prickly Paradigm Press, 2003).

—*Crystals, Fabrics, and Fields: Metaphors That Shape Embryos* (Berkeley, CA: North Atlantic, 2003).

—*When Species Meet* (Minneapolis, MN: Minnesota University Press, 2008).

Harmon, Katherine, *The Map as Art: Contemporary Artists Explore Cartography* (Princeton, NJ: Princeton Architectural Press, 2010).

Harper, Glen (ed.), *Interventions and Provocations: Conversations in Art, Culture, and Resistance* (Albany, NY: SUNY Press, 1998).

Harrison, Helen Mayer and Newton Harrison, *Shrimp Farm, Survival Piece #2*. Available at http://goo.gl/2xzuHw (accessed 10 December 2014).

Hastie, Amelie, 'Senseless eating', *parallax* 66 (2013), pp. 64–73.

Hauser, Jens, *L'art biotech* (Nantes: Le Lieu Unique, 2003).

—'Taxonomy of an etymological monster', in Gerfried Stocker and Christine Schopf (eds), *Ars Electronica: Hybrid – Living in Paradox* (Ostfildern-Ruit: Hatje Cantz, 2005), pp. 182–93.

—'Bio art – taxonomy of an etymological monster', *Hybrid: Living in Paradise* (Ars Electronica, 2005). Available at http://90.146.8.18/en/archives/festival_archive/festival_catalogs/festival_artikel.asp?iProjectID=13286 (accessed 30 January 2014).

Hawkins, Joan, 'When taste politics meet terror: The Critical Art Ensemble on trial', *1000 Days of Theory* (2005). Available at http://www.ctheory.net/articles.aspx?id=482 (accessed 18 February 2015).

Hayles, Katherine, *How We Became Posthuman: Virtual Bodies in Cybernetics, Literature, and Informatics* (Chicago, IL: Chicago University Press, 1999).

Hayward, Eva, 'FingeryEyes: Impressions of cup corals', *Cultural Anthropology* 25/4 (2010), pp. 577–99.

Heiferman, Marvin, Carole Kismaric and Ian Berry, *Paradise Now: Picturing The Genetic Revolution* (Saratoga Springs, NY: Tang Teaching Museum, 2001).

Heinzelmann, Ursula, 'An interview with René Redzepi: Noma, Copenhagen', *Gastronomica: The Journal of Food and Culture* 10/3 (2010), pp. 97–101.

Heldke, Lisa, 'Recipes for theory making', *Hypatia* 3/2 (Summer 1988), pp. 15–29.

Henke, Christopher and Thomas Gieryn, 'Sites of scientific practice: the enduring importance of place', in Edward Hackett, Olga Amsterdamska, Michael Lynch and Judy Wajcman (eds), *The Handbook of Science and Technology Studies*, 3rd edn (Cambridge, MA: MIT Press, 2008), pp. 353–76.

Heselmans, Marianne, 'Cultivated meat (NRC)', *New Harvest*, 10 September 2005. Available at http://www.new-harvest.org/2005/09/cultivated-meat-nrc/ (accessed 30 January 2014).

Higgins, Hannah, *FLUXUS Experience* (Berkeley, CA: University of California Press, 2002).

High, Kathy, *Embracing Animal* (2004). Available at http://www.embracinganimal.com/ (accessed 15 September 2012).

Hirsch, Robert, 'The strange case of Steve Kurtz: Critical Art Ensemble and the price of freedom' *Afterimage* 32/6 (2005), pp. 22–32.

Ho, Mae-Wan and Lim Li Ching, *GMO Free: Exposing the Hazards of Biotechnology to Ensure the Integrity of Our Food Supply* (Ridgefield, CT: Vital Health, 2004).

Hollows, Joanne, 'The feminist and the cook: Julia Child, Betty Friedan and Domestic Femininity', in Lydia Martens and Emma Casey (eds), *Gender and Consumption: Domestic Cultures and the Commercialisation of Everyday Life* (Farnham: Ashgate, 2007), pp. 33–48.

'Home Economics graduates'. Available at http://rmc.library.cornell.edu/homeEc/3careers/chart.html (accessed 9 March 2014).

Horner, Jennifer R., '*Betty Crocker's Picture Cookbook*: a gendered ritual response to social crises of the postwar era', *Journal of Communication Inquiry* 24 (2000), pp. 332–345.

Hovey, Kathryn, *Anarchy and Community in the New American West* (Albuquerque, NM: University of New Mexico Press, 2005).

Howard, Ted and Jeremy Rifkin, *Who Should Play God? The Artificial Creation of Life and what it Means for the Future of the Human Race* (New York: Delacorte, 1977).

Hsiao, F.S., W.S. Lian, S.P. Lin, C.J. Lin, Y.S. Lin et al., 'Toward an ideal animal model to trace donor cell fates after stem cell therapy: production of stably labeled multipotent mesenchymal stem cells from bone marrow of transgenic pigs harboring enhanced green fluorescence protein gene', 24 June 2011. Available at http://www.ncbi.nlm.nih.gov/pubmed/21705633 (accessed 15 September 2012).

Hurd, Emma, 'UN pleads for £124m to help starving Niger', *Sky News*, 30 April 2010. Available at http://news.sky.com/story/775823/un-pleads-for-124m-to-help-starving-niger (accessed 20 January 2015).

Huxley, Julian, *Memories* (New York: Harper & Row, 1970).

Infield, Glenn 'A computer in the *basement*?', *Popular Mechanics* (April 1968), pp. 77–9, 209, 229.

Jardin, Xeni, 'BBtv – Star Simpson's fuzzy logic, MacGyver, MIT lasers, and trippy glasses: Maker Faire with Phil Torrone', 8 May 2008. Available at http://www.boingboing.net/2008/05/08/bbtv-star-simpsons-f.html (accessed 4 February 2014).

Jaworski, Birdie, *My Tiny Vegas: Life between the Sangre de Christo Mountains and the Great Plains* (Create Space, 2009).

Jeremijenko, Natalie, *X-Species: The Adventure Club* brochure. Available at http://www.environmentalhealthclinic.net/xooz/wp-content/uploads/2010/08/100721_xspecies-adventureclub.pdf (accessed 13 February 2015).

—and Heath Bunting, *Biotech Hobbyist Magazine*, n.d. Available at http://www.nyu.edu/projects/xdesign/biotechhobbyist/skin.html (accessed 30 January 2014).

Johnson, Matthew, *GMO Free Diet: How to Stay Healthy by Identifying and Avoiding Dangerous Foods* (Digital Direct Publishing, 2013).

Jones, Caroline and Peter Galison (eds), *Picturing Science Producing Art* (New York: Routledge, 1998).

Kac, Eduardo, *Telepresence and Bio Art: Networking Humans, Rabbits and Robots* (Ann Arbor, MI: Michigan University Press, 2005).

—(ed.), *Signs of Life: Bioart and Beyond* (Cambridge, MA: MIT Press, 2007).

—*Cypher*, n.d. Available at http://www.ekac.org/cypher.text.html (accessed 30 January 2014).

—*GFP Bunny*, n.d., nn. 17 and 18. Available at http://www.ekac.org/gfpbunny.html#gfpbunnyanchor (accessed 15 September 2012).

Kane, Stephanie and Pauline Greenhill, 'A feminist perspective on bioterror: from anthrax to the critical art ensemble', *Signs: Journal of Women in Culture and Society* 33/1 (2007), pp. 53–80.

Katyal, Sonia, 'Semiotic disobedience', *Washington University Law Review* 84/2 (2006), pp. 489–571.

Kelley, Lindsay, 'Plumpiñon', in Eben Kirksey (ed.), *The Multispecies Salon* (Chapel Hill, NC: Duke University Press, 2014), pp. 122–134.

—'Tranimals', *Transgender Studies Quarterly* 1/1–2 (May 2014), pp. 226–8.

—'Digesting wetlands: cooking and eating across species with Natalie Jeremijenko's *Cross(x)Species Adventure Club*', forthcoming in Silvia Bottinelli and Margherita d'Ayala (eds), *What's Cooking? Food, Art and Counterculture* (Fayetteville, AR: University of Arkansas Press, 2015).

—and Eva Hayward, "Carnal light: following the white rabbit', *parallax* 66, 'bon appetit' (January 2013), pp. 114–127.

—and Lynn Turner (guest eds), *parallax* 66: 'bon appétit' 19/1 (2013).

Kershner, Bruce and Craig Tufts, *National Wildlife Federation Field Guide to Trees of North America* (New York: Sterling, 2008).

Kester, Grant, *Conversation Pieces: Community and Communication in Modern Art* (Berkeley, CA: Berkeley University Press, 2004).

—*Groundworks: Environmental Collaboration in Contemporary Art* (Pittsburgh, PA: Carnagie Mellon University Press, 2005).

Kirksey, Eben (ed.), *The Multispecies Salon* (Chapel Hill, NC: Duke University Press, 2014).

Knorr-Cetina, Karin and Michael Mulkay (eds), *Science Observed: Perspectives on the Social Study of Science* (Beverly Hills, CA: Sage, 1983).

Krauss, Rosalind, Annette Michelson, George Baker, Yve-Alain Bois, Benjamin H.D. Buchloh, Hal Foster, Denis Hollier, Carrie Lambert-Beatty, Mignon Nixon and Malcolm Turvey (eds), 'In what ways have artists, academics, and cultural institutions responded to the US-led invasion and occupation of Iraq?', *OCTOBER* 123 (Winter 2008).

Lacy, Suzanne (ed.), *Mapping the Terrain: New Genre Public Art* (Seattle, WA: Bay Press, 1995).

Landecker, Hannah, 'Immortality, *in vitro*: a history of the HeLa cell line', in Paul Brodwin (ed.), *Biotechnology and Culture: Bodies, Anxieties, Ethics* (Bloomington, IN: Indiana University Press, 2000), pp. 53–74.

—'Cellular Features: microcinematography and early film theory', *Critical Inquiry* 3/4 (2005), pp. 903–37.

—'Living differently in time: plasticity, temporality, and cellular biotechnologies', *Culture Machine* 7 (2005). Available at http://www.culturemachine.net/index.php/cm/article/view/26/33 (accessed 30 January 2014).

—*Culturing Life: How Cells Became Technologies* (Cambridge, MA: Harvard University Press, 2007).

Lanner, Ronald, *The Piñon Pine: A Natural and Cultural History* (Reno, NV: University of Nevada Press, 1981).

Latour, Bruno, 'Give me a laboratory and I will raise the world', in Karin Knorr-Cetina and Michael Mulkay (eds), *Science Observed: Perspectives on the Social Study of Science* (Beverly Hills, CA: Sage, 1983), pp. 141–70.

—*Science in Action: How to Follow Scientists and Engineers through Society* (Cambridge, MA: Harvard University Press, 1987).

—'How to be iconophilic in art, science, and religion?', in Caroline Jones and Peter Galison (eds), *Picturing Science Producing Art* (New York: Routledge, 1998), pp. 418–40.

—'From realpolitik to dingpolitik or how to make things public', in Bruno Latour and Peter Weibel (eds), *Making Things Public: Atmospheres of Democracy* (Cambridge, MA: MIT Press, 2005), pp. 4–31.

—and Steve Woolgar, *Laboratory Life: The Construction of Scientific Facts* (Princeton, NJ: Princeton University Press, 1986).

Lee, Linda, 'Bridal hunger games: losing weight in time for the wedding', *New York Times*, 13 April 2012. Available at http://www.nytimes.com/2012/04/15/fashion/weddings/Losing-Weight-in-Time-for-the-Wedding.html?_r=0 (accessed 19 February 2015).

Light, Jennifer, 'When computers were women', *Technology and Culture* 40/3 (1999), pp. 455–83.

Lingis, Alphonso, *Phenomenological Explanations* (Dordrect: Martinus Nijhoff, 1986).

Lippard, Lucy, *Weather Report: Art and Climate Change* (Boulder, CO: Boulder Museum of Contemporary Art, 2007).

Lok See Fu, Joyce, 'The potential decline of artistic activity in the wake of the Patriot Act: the case surrounding Steven Kurtz and the Critical Art Ensemble', *The Columbia Journal of Law and the Arts* 29/1 (Fall 2005), pp. 83–106.

Lupton, Ellen, *Mechanical Brides: Women and Machines from Home to Office* (New York: Cooper Hewitt, 1993).

Lushetich, Natasha, 'Hovering between being and non-being: Nishidian interexpression and the Fluxkit', *Performance Research: A Journal of the Performing Arts* 16/4 (2011), pp. 73–9.

Malinin, Theodore, *Surgery and Life: The Extraordinary Career of Alexis Carrel* (New York: Harcourt Brace Jovanovich, 1979).

Marder, Michael, *Plant-Thinking: Toward a Philosophy of Vegetal Life* (New York: Columbia, 2013).

Marks, Susan, *Finding Betty Crocker: The Secret Life of America's First Lady of Food* (New York: Simon & Schuster, 2005).

Massachusetts Board of Health, *Massachusetts Board of Health Guidebook*. Available at http://www.mahb.org/Library/Guide%20book/gbook10.pdf (accessed 13 February 2015).

Mayes, Vernon and Barbara Bayless Lacy, *Nanise': A Navajo Herbal* (Tsaile, AZ: Navajo Community College Press, 1989).

McBride, Anne E., 'Have your Coke and eat it too: what cooking with Coca-Cola says about cultural imperialism', *Gastronomica: The Journal of Food and Culture* 5/1 (2005), pp. 80–7.

McFeely, Mary Drake, *Can She Bake a Cherry Pie? American Women and the Kitchen in the Twentieth Century* (Amherst, MA: Massachusetts University Press, 2000).

McGee, Harold, *On Food and Cooking: The Science and Lore of the Kitchen* (New York: Scribner, 2004).

McLain, Rebecca, 'Management guidelines for expanding pinyon nut production in Colorado's pinyon-juniper woodlands', Institute for Culture and Ecology (2008).

McLuhan, Marshall, *The Mechanical Bride: Folklore of Industrial Man* (Corte Madera, CA: Ginko Press, 2002).

'Meet animal meat', Available at http://www.genna.gender.uu.se/conferences-events/conferences-workshops/meet-animal-meat/ (accessed 7 February 2014).

'Men in black', *Time*, 13 June, 1938, pp. 40–3.

Merchant, Carolyn, *Earthcare: Women and the Environment* (New York: Routledge, 1995).

Metcalf, Jacob, 'Meet shmeat: food system ethics, biotechnology and re-worlding technoscience', *parallax* 19/1 (2013), pp. 74–87.

Mintz, Sidney, *Tasting Food, Tasting Freedom: Excursions into Eating, Culture, and the Past* (Boston, MA: Beacon, 1996).

Mirzoeff, Nicholas, 'Visualizing the anthropocene', *Public Culture* 26/2 (2014), pp. 213–32.

Mitchell, J. William, Alan S. Inouye, and Marjory S. Blumenthal (eds), *Beyond Productivity: Information Technology, Innovation, and Creativity* (Washington DC: National Academies Press, 2003).

Mitchell, Robert, *Bioart and the Vitality of Media* (Seattle, WA: Washington University Press, 2010).

Mitchell, W.J.T., *Picture Theory* (Chicago, IL: Chicago University Press, 1994).

—'The work of art in the age of biocybernetic reproduction', *Modernism/Modernity* 10/3 (2003), pp. 481–500.

Montano, Linda, 'Food & art', *High Performance* 4 (Winter 1981–2), pp. 45–54.

—*Performance Artists Talking in the Eighties* (Berkeley, CA: University of California, 2000).

Morse, Margaret, *Virtualities: Television, Media Art, and Cyberculture* (Bloomington, IN: Indiana University Press, 1998).

Morton, Timothy, *Ecology Without Nature: Rethinking Environmental Aesthetics* (Cambridge, MA: Harvard University Press, 2007).

Mulvey, Laura, 'Visual pleasure and narrative cinema', *Screen* 16/3 (1975), pp. 6–18.

Munster, Anna, 'Why is bioart *not* terrorism?: some critical nodes in the networks of infomatice life', *Culture Machine* 7 (2005). Available at http://www.culturemachine.net/index.php/cm/article/view/31/38 (accessed 3 February 2014).

—*An Aesthesia of Networks* (Cambridge, MA: MIT Press, 2013).

Nadis, Steve, 'Science for art's sake', *Nature* 407 (2000), pp. 668–70.

National Research Council, *The Lost Crops of Africa, Volume III: Fruits* (Washington DC: National Academies Press, 2008).

Nayar, Pramod, 'The new monstrous: digital bodies, genomic arts and aesthetics', *Nebula* 4/2 (2007), pp. 1–19.

Neuhaus, Jessamyn, *Manly Meals and Mom's Home Cooking: Cookbooks and Gender in Modern America* (Baltimore, MD: Johns Hopkins University Press, 2003).

Next Nature Network, *The In Vitro Meat Cookbook* (Amsterdam: BIS Publishers, 2014).

'Nintendo Wii controlled smarthome', *Liquidice's Nintendo Wii Hacks*, 23 December 2006. Available at http://wiihacks.blogspot.co.uk/2006/12/nintendo-wii-controlled-smarthome.html (accessed 23 January 2014).

Norton, Bryan, *Searching for Sustainability: Interdisciplinary Essays in the Philosophy of Conservation Biology* (Cambridge: Cambridge University Press, 2003).

Obrist, Hans Ulrich and Barbara Vanderlinden (eds), *Laboratorium* (Antwerp: Dumont, 1999).

—and Kate Fowlie (eds), *Do It: The Compendium* (New York: Distributed Art Publishers, 2013).

Olson Donald, Marilynn S. Olson, Russell L. Doescher, Lance L. Lambert, David E. Lemke, Angela M. Carl, Ross Johnson, Sandra D. Smith and Kent H. Trede, 'Piñon Pines and the Route of Cabeza de Vaca', *Southwest Historical Quarterly* 101 (1997), pp. 174–86.

Opel, Andy, Joseé Johnston and Richard Wilk (eds), *Environmental Communication: A Journal of Nature and Culture*: special issue 'Food, culture and the environment: communicating about what we eat' 4/3 (2010).

O'Rourke, Karen, *Walking and Mapping: Artists as Cartographers* (Cambridge, MA: MIT Press, 2013).

Paris, Jeanne, *Your Future as a Home Economist* (New York: Richards Rosen, 1964).

Paxson, Heather, 'Post-Pasteurian cultures: the microbiopolitics of raw-milk cheese in the United States', *Cultural Anthropology* 23/1 (2008), pp. 15–47.

Pendergrast, Mark, *For God, Country and Coca-Cola: The Definitive History of the Great American Soft Drink and the Company that Makes It* (New York: Basic Books, 1993).

Pentecost, Claire, 'What did you eat and when did you know it?', *Art Journal* 61/3 (Autumn 2002), pp. 47–62.

—'Reaching the limit', in Beatriz da Costa and Kavita Philip (eds), *Tactical Biopolitics* (Cambridge, MA: MIT Press, 2008), pp. 365–385.

People for the Ethical Treatment of Animals (PETA), 'In vitro meat prize deadline extended', 25 June 2012. Available at http://www.peta.org/blog/vitro-meat-prize-deadline-extended/ (accessed 29 January 2014).

—'About PETA'. Available at http://www.peta.org/about-peta/ (accessed 7 February 2014).

—'PETA's "In vitro" chicken contest', n.d. Available at http://www.peta.org/features/vitro-meat-contest/ (accessed 18 April 2014).

Pereval, Verena, 'Architecture of a demonstration: building facts, walls and publics', 4S & EASST Conference, 2004. Abstract available at http://www.csi.ensmp.fr/WebCSI/4S/abstract/abstract_papier.php?numero=36 (accessed 30 January 2014).

Petrinovich, Lewis, *The Cannibal Within* (New York: Aldine de Gruyter, 2000).

Phelan, Mary C., 'Techniques for mammalian cell tissue culture', *Current Protocols in Cytometry*, 36:A.3B.1–A.3B.18, 1 May 2006. Available at http://onlinelibrary.wiley.com/doi/10.1002/0471142956.cya03bs36/full#cya03b-rec-0001 (accessed 30 January 2014).

Philipkoski, Kristen, 'RIP: Alba, the glowing bunny', *WIRED*, 12 August 2002. Available at http://www.wired.com/medtech/health/news/2002/08/54399 (accessed 15 September 2012).

Phillips, Velma, *Home Economics Careers for You* (New York: Harper & Brothers, 1962).

Pickering, Andrew, *The Mangle of Practice* (Chicago, IL: Chicago University Press, 1995).

Pineau, Courtney, *The Non-GMO Cookbook: Recipes and Advice for a Non-GMO Lifestyle* (New York: Skyhorse, 2013).

Pollan, Michael, *The Omnivore's Dilemma: A Natural History of Four Meals* (New York: Penguin, 2006).

—*In Defense of Food: An Eater's Manifesto* (New York: Penguin, 2008).

Powell, Jody, *The Other Side of the Story* (New York: Morrow, 1984).

Pratt, Mary Louise, *Imperial Eyes: Travel Writing and Transculturation* (New York: Routledge, 2008).

Puncer, Mojca, 'Artistic research on life forms: exploring the intersections of science, art and life in the context of globalization', *Leonardo* 41/5 (2008), pp. 468–77.

Querling, Jonathan, 'Resistance takes root: protests against genetically modified crops', *The Ecologist*, December 2000, p. 57.

Rabbit Welfare Association and Fund, 'Breeds: choosing your perfect pet rabbit', n.d. Available at http://www.rabbitwelfare.co.uk/resources/?section=breeds.html (accessed 15 September 2012).

Reggiani, Andrés Horacio, *God's Eugenicist: Alexis Carrel and the Sociobiology of Decline* (New York: Berghahn, 2007).

Reichle, Ingeborg, *Art in the Age of Technoscience: Genetic Engineering, Robotics, and Artificial Life in Contemporary Art* (New York: Springer, 2009).

Renova Life Inc., 'Products and services|GFP rabbits', 15 January 2014. Available at http://www.renovalife.com/gfp_rabbits.php (accessed 15 September 2012).

Rhode, David, *Native Plants of Southern Nevada: An Ethnobotany* (Salt Lake City, UT: University of Utah Press, 2002).

Richards, Ellen, *The Chemistry of Cooking and Cleaning* (Boston, MA: Estes & Lauriat, 1882).

—'Report of the Massachusetts Board of World's Fair managers' (Boston, 1983). Available at http://libraries.mit.edu/archives/exhibits/esr/esr-rumford.html (accessed 19 January 2015).

Ricoeur, Paul, *Rule of Metaphor: The Creation of Meaning in Language* (New York: Routledge, 2003).

Rifkin, Jeremy, *Beyond Beef: The Rise and Fall of the Cattle Culture* (New York: Dutton, 1992).

Riss, Jorjo (ed.), *Good Food Gone Bad: Genetically Modified Recipes for Disaster* (Brussels: Greenpeace, 2010). Available at http://www.greenpeace.org/eu-unit/en/Publications/2010/GMO-cookbook/ and http://www.greenpeace.org/eu-unit/Global/eu-unit/reports-briefings/2010/4/GMO-cookbook.pdf (accessed 18 January 2014).

Roosth, Sophia, 'Of foams and formalisms: scientific expertise and craft practice in molecular gastronomy', *American Anthropologist* 115/1 (2013), pp. 4–16.

Rosebraugh, Craig, *Burning Rage of a Dying Planet: Speaking for the Earth Liberation Front* (New York: Lantern Books, 2004).

Rosler, Martha, 'Romances of the meal', in Hans Ulrich Obrist and Barbara Vanderlinden, *Laboratorium* (Antwerp: Dumont, 1999).

—*Decoys and Disruptions: Selected Writings 1975–2001* (Cambridge, MA: MIT Press, 2004).

Ross, Phil, *Biotechnique* (San Francisco, CA: Yerba Buena Center for the Arts, 2007).

Rossiter, Margaret, *Women Scientists in America: Struggles and Strategies to 1940* (Baltimore, MD: Johns Hopkins University Press, 1982).

—*Women Scientists in America: Before Affirmative Action, 1940–1972* (Baltimore: Johns Hopkins University Press, 1995).

Rothfels, Nigel (ed.), *Representing Animals* (Bloomington, IN: Indiana University Press, 2002).

'The Rumford Kitchen exhibit at World's Columbian Exposition, Chicago, 1893: excerpts from *Report of the Massachusetts Board of World's Fair Managers*'.

Available at http://libraries.mit.edu/archives/exhibits/esr/esr-rumford.html (accessed 9 March 2014).

Ryan, Leslie, 'Art + ecology: land reclamation works of artists Robert Smithson, Robert Morris, and Helen Mayer Harrison and Newton Harrison', *Environmental Philosophy* 4/1 and 2 (2007), pp. 95–116.

Ryan, Zoë and Mel Chin, 'A conversation with Mel Chin', *Log* 8 (Summer 2006), pp. 59–68.

Sagan, Dorion, 'Introduction: umwelt after Uexküll', in Jakob von Uexküll, *A Foray into the Worlds of Animals and Humans: With A Theory of Meaning* (Minneapolis, MN: Minnesota University Press, 2010), pp. 1–34.

Savinelli, Alfred, *Plants of Power: Native American Ceremony and the Use of Sacred Plants* (Summertown, TN: Native Voices, 2002).

Schneider, Rebecca, 'Nomadmedia: On Critical Art Ensemble', *The Drama Review* 44/4 (2000), pp. 120–31.

—and Jon McKenzie, 'Keep your eyes on the front and watch your back', *The Drama Review* 48/4 (2004), pp. 5–10.

Schor, Mira, *A Decade of Negative Thinking: Essays on Art, Politics, and Daily Life* (Durham, NC: Duke University Press, 2009).

Schurman, Rachel and Dennis Doyle Takahashi Kelso (eds), *Engineering Trouble: Biotechnology and Its Discontents* (Berkeley, CA: University of California Press, 2003).

Schwartz, John, 'PETA's latest tactic: $1 million for fake meat', *New York Times*, 21 April 2008. Available at http://www.nytimes.com/2008/04/21/us/21meat.html (accessed 29 January 2014).

Senior, Adele, 'Towards a (semi-)discourse of the semi-living; the undecidability of a life exposed to death', *Technoetic Arts: A Journal of Speculative Research* 5/2 (2007), pp. 97–112.

Severson, Kim, 'Don't tell the kids', 3 March 2010. Available at http://query.nytimes.com/gst/fullpage.html?res=9905E1D8103AF930A35750C0A9669D8B63&pagewanted=all (accessed 15 September 2012).

Shapin, Steven and Simon Schaeffer, *Leviathan and the Air-Pump: Hobbes, Boyle, and the Experimental Life* (Princeton, NJ: Princeton University Press, 1985).

Shapiro, Laura, '"I guarantee": Betty Crocker and the woman in the kitchen', in Arlene Voski Avakian and Barbara Haber (eds), *From Betty Crocker to Feminist Food Studies: Critical Perspectives on Women and Food* (Boston, MA: Massachusetts University Press, 2005), pp. 29–40.

Sherman, H.C., J.C. Winters, and V. Phillips, 'Efficiency of oat protein in adult human nutrition', *The Journal of Biological Chemistry* 39/1 (1919), pp. 53–62.

Sholette, Gregory, 'Disciplining the avant-garde: The United States versus the Critical Art Ensemble', *Circa* 112 (Summer 2005), pp. 50–9.

Shukin, Nicole, *Animal Capital: Rendering Life in Biopolitical Times* (Minneapolis, MN: Minnesota University Press, 2009).

Smith, Julie Ann, 'Beyond dominance and affection: living with rabbits in post-humanist households', *Society and Animals* 11/2 (2003), pp. 181–97.

Smith, Stephanie, *Feast: Radical Hospitality in Contemporary Art* (Chicago, IL: Smart Museum of Art, 2013).

—and Victor Margolin (eds), *Beyond Green: Towards a Sustainable Art* (Chicago, IL: Smart Museum of Art, 2009).

Snyder, Midori, 'Madrid's Illiad: the list of miners' names: Part III.' Available at http://msnyder.typepad.com/the_labyrinth/2011/01/madrids-illiad-the-list-of-names.html (accessed 30 January 2015).

Sonfist, Alan (ed.), *Art in the Land: A Critical Anthology of Environmental Art* (New York: Plume, 1983).

Spector, Tami, 'Nanoaesthetics: from the molecular to the machine', *Representations* 117/1 (Winter 2012), pp. 1–29.

Spicer, Dag, 'If you can't stand the coding, stay out of the kitchen: three chapters in the history of home automation', *Dr Dobbs' Portal*, 12 August 2000. Available at http://www.ddj.com/184404040 (accessed 23 January 2014).

Squier, Susan, 'Life and death at Strangeways: the tissue culture point of view', in Paul Brodwin (ed.), *Biotechnology and Culture: Bodies, Anxieties, Ethics* (Bloomington, IN: Indiana University Press, 2000), pp. 27–52.

Stadler, Beda, *Gene an die Gabel* (Zurich: Internutrition, 2001).

Stage, Sarah, 'Ellen Richards and the social significance of the Home Economics movement', in Sarah Stage and Virginia B. Vincenti (eds), *Rethinking Home Economics: Women and the History of a Profession* (Ithaca, NY: Cornell University Press, 1997), pp. 17–33.

—and Virginia B. Vincenti (eds), *Rethinking Home Economics: Women and the History of a Profession* (Ithaca, NY: Cornell University Press, 1997).

Stengers, Isabel, *Cosmopolitics II*, trans. Robert Bononno (Minneapolis, MN: Minnesota University Press, 2011).

Stewart, Alastair, 'Millions face starvation as Niger prays in vain for rain', *The Independent*, 9 July 2010. Available at http://www.independent.co.uk/news/world/africa/millions-face-starvation-as-niger-prays-in-vain-for-rain-2022190.html (accessed 20 January 2015).

Stimson, Blake and Gregory Sholette (eds), *Collectivism after Modernism: The Art of Social Imagination after 1945* (Minneapolis, MN: Minnesota University Press, 2007).

Stock, Timothy and Warren Heise, 'Suspect culture: an exposition of the death of Hope Kurtz', in John Knechtel (ed.), *Suspect* (Cambridge, MA: MIT Press, 2006), pp. 332–49.

Stocker, Gerfried and Christine Schopf (eds), *Hybrid: Living in Paradox* (Linz: Ars Electronica, 2005).

Stokstad, Erik, 'Engineered fish: friend or foe of the environment?' *Science* 297/5588 (13 September 2002), pp. 1797–9.

Stracey, Frances, 'Bio-art: the ethics behind the aesthetics', *Nature Reviews Molecular Cell Biology* 10 (2009), pp. 496–500.

Strasser, Susan, *Never Done: A History of American Housework* (New York: Holt, 1982).

Strom, Stephanie, 'Group seeks special label for food: "natural"', *New York Times*, 19 December 2013. Available at http://www.nytimes.com/2013/12/20/business/trade-group-seeks-natural-label-on-modified-food.html?_r=0 (accessed 24 April 2014).

Swallow, Pamela C., *The Remarkable Life and Career of Ellen Swallow Richards: Pioneer in Science and Technology* (New York: Wiley, 2014).

Swinbank, Vicki, 'The sexual politics of cooking: a feminist analysis of culinary hierarchy in Western culture', *Journal of Historical Sociology* 15/4 (2002), pp. 464–94.

Szerszynski, Bronislaw, 'The post-ecologist condition: irony as symptom and cure', *Environmental Politics* 16/2 (2007), pp. 337–55.

Tabatabai Asbaghi, Pandora and Jerry Gorovoy, *Louise Bourgeois: Blue Days and Pink Days* (Milan: Fondazione Prada, 1997).

Tannahill, Ray, *Flesh and Blood: The History of the Cannibal Complex* (Boston, MA: Little, Brown, 1996).

Thacker, Eugene, 'Data made flesh: biotechnology and the discourse of the posthuman', *Cultural Critique* 53 (Winter 2003), pp. 72–97.

—*The Global Genome: Biotechnology, Politics and Culture* (Cambridge, MA: MIT Press, 2005).

'The Ultimate Dallas Salad', *D Magazine* (September 2013), p. 55.

Theophano, Janet, *Eat My Words: Reading Women's Lives through the Cookbooks They Wrote* (New York: Palgrave, 2002).

This, Hervé, *Molecular Gastronomy: Exploring the Science of Flavor* (New York: Columbia University Press, 2005).

Thomas, Paul, *Nanoart: The Immateriality of Art* (Chicago, IL: Intellect, 2013).

Thompson, Carolyn, 'Prof sent bacteria to artist; no jail', Associated Press, 12 January 2008. Available at http://www.usatoday.com/news/nation/2008-02-12-3296235700_x.htm (accessed 3 February 2014).

Thompson, Nato and Gregory Sholette with Joseph Thompson, Nicholas Mirzoeff and Ondine C. Chavoya (eds), *The Interventionists: Users' Manual for the Creative Disruption of Everyday Life* (Cambridge, MA: MIT Press, 2004).

—with Joseph Thompson, Christoph Cox (eds), *Becoming Animal: Contemporary Art in the Animal Kingdom* (North Adams, MA: MASS MoCA Publications, 2005).

Thompson, Patricia, *Bringing Feminism Home: Home Economics and the Hestian Connection* (Charlottetown, PEI: Home Economics Publishing Collective, 1992).

Tiravanija, Rirkrit, *Soccer Half-Time Cookery Book* (Nurnberg: Verlag Fur Moderne Kunst, 2006).

—and Francesca Grassi (eds), *Rirkrit Tiravanija: A Retrospective (Tomorrow Is Another Fine Day)* (Miami, FL: Bridge House, 2007).

Tissue Culture and Art Project, 'Semi-living food – "disembodied cuisine"'. Available at http://www.tca.uwa.edu.au/disembodied/dis_text.html (accessed 30 January 2014).

—'TC&A biographies of Project members'. Available at http://www.tca.uwa.edu.au/atGlance/bios.html (accessed 30 January 2014).

Tomasula, Steve, 'Genetic art and the aesthetics of biology', *Leonardo* 35/2 (2002), pp. 137–44.

Tomayko, James, 'Electronic Computer for Home Operation (ECHO): the first home computer', *IEEE Annals of the History of Computing* 16/3 (1994), pp. 59–61.

Triscott, Nicola, 'Performative science: the case of Critical Art Ensemble', in Maria Chatzichristodoulou, Janis Jeffries and Rachel Zerihan (eds), *Interfaces of Performance* (London: Ashgate, 2009), pp. 153–66.

Trivedi, Bijal P., 'Introducing ANDi: The first genetically modified monkey', *Genome News Network*, 16 January 2001. Available at http://www.genomenewsnetwork.org/articles/01_01/ANDi.shtml (accessed 15 September 2012).

Tuchman, Maurice, *A Report on the Art and Technology Program of the Los Angeles County Museum of Art* (Los Angeles, CA: LACMA, 1971). Available at https://archive.org/stream/reportonarttechn00losa_/reportonarttechn00losa#page/n0/mode/2up (accessed 22 January, 2015).

Tuosto, Kylie, 'The "grunt truth" of embedded journalism: the new media/military relationship', *Stanford Journal of International Relations* 20 (Fall/Winter 2008), pp. 20–31.

Twilley, Nicola, 'Cross-species dining: an interview with Natalie Jeremijenko and Mihir Desai', *Edible Geography*, October 27, 2010. Available at http://www.ediblegeography.com/cross-species-dining-an-interview-with-natalie-jeremijenko-and-mihir-desai/ (accessed 13 February 2015).

United States Census Bureau, American Fact Finder, 'Hispanic or Latino by Type: 2010 Census Summary File 1', *American Fact Finder*. Available at http://factfinder2.census.gov/faces/tableservices/jsf/pages/productview.xhtml?pid=DEC_10_SF1_QTP10&prodType=table (accessed 30 January 2015).

United States Department of Agriculture and Office of Education, Department of Health, Education, and Welfare, *Titles of Completed Theses in Home Economics and Related Fields in Colleges and Universities of the United States, 1954–1955* (Washington DC: US Government Printing Office, 1955).

United States Department of Labor, Women's Bureau, 'Nontraditional occupations for women in 2009', April 2010. Available at http://goo.gl/gFwrUy (accessed 14 January 2014).

United States Environmental Protection Agency, 'How people are exposed to mercury', n.d. Available at http://goo.gl/fOaCHX (accessed 23 May 2014).

United States War Department, *You're Going to Employ Women* (Washington DC: US Government Printing Office, 1943).

US Food and Drug Administration (FDA), 'FDA Basics: What is the meaning of "natural" on the label of food?', n.d. Available at http://www.fda.gov/aboutfda/transparency/basics/ucm214868.htm (accessed 24 April 2014).

Vincenti, Virginia B., 'Chronology of events and movements which have defined and shaped Home Economics', in Sarah Stage and Virginia B. Vincenti (eds), *Rethinking Home Economics: Women and the History of a Profession* (Ithaca, NY: Cornell University Press, 1997), pp. 321–30.

Vint, Sheryl, 'Disembodied Cuisine: Eating Well in The Mad and Oryx and Crake', SLSA 2009. Available at http://www.litsciarts.org/slsa09/program.php (accessed 31 August 2015).

Vita-More, Natasha, 'Brave bioart 2: shedding the bio, amassing the nano, and cultivating posthuman life', *Technoetic Arts: A Journal of Speculative Research* 5/3 (2007), pp. 171–86.

Wajcman, Judy, *Feminism Confronts Technology* (University Park, PA: Pennsylvania State University Press, 1991).

Walton, Priscilla, *Our Cannibals, Ourselves* (Urbana, IL: Illinois University Press, 2004).

Ward, Paula Reed, 'Doctor guilty of mailing bacteria for use in artwork', *Pittsburgh Post-Gazette*, 12 October 2007. Available at http://www.post-gazette.com/pg/07285/824828-85.stm (accessed 3 February 2014).

Ward, William, 'Experiments with GFP – the art and the science', n.d. Available at http://dbm.rutgers.edu/11115433SpecialTopicsBiochem.html (accessed 30 January 2015).

Watts, Oliver, 'The image and the terrorist', *Law Text Culture* 10 (2006), pp. 221–38.

Weibel, Peter and Ljiljana Fruk (eds), *Molecular Aesthetics* (Cambridge, MA: MIT Press, 2013).

Weintraub, Linda, *To Life!: Eco Art in Pursuit of a Sustainable Planet* (Berkeley, CA: University of California Press, 2012).

Weiser, Kathy, 'New Mexico legends: Madrid – a ghost town reborn', *Legends of America*, September 2011. Available at http://www.legendsofamerica.com/nm-madrid.html (accessed 29 January 2015).

Wells, Herbert George, *The Sleeper Awakes* (Lincoln, NE: Nebraska University Press 2000).

'What were practice apartments?' Available at http://rmc.library.cornell.edu/homeEc/cases/apartments.html (accessed 9 March 2014).

Whirlpool, 'Smart appliances featuring smart home technology', n.d. Available at http://www.whirlpool.com/smart-appliances/ (accessed 23 January 2014).

Whitehouse, David, 'GM fish glows in the bowl', *BBC News*, 27 June 2003. Available at http://news.bbc.co.uk/2/hi/science/nature/3026104.stm (accessed 15 September 2012).

Willet, Jennifer, 'Part 1: GFP LAB', *Bioart: Contemporary Art and the Life Sciences*, 13 February 2012. Available at http://bioartwindsor.blogspot.com/2012/02/part-1-gfp-lab.html (accessed 15 September 2012).

Wines, Michael, 'Hope for hungry children, arriving in a foil Packet', *New York Times*, 8 August 2005. Available at http://www.nytimes.com/2005/08/08/international/africa/08niger.html (accessed 25 September 2014).

Wolfe, Cary (ed.), *Zoontologies: The Question of the Animal* (Minneapolis, MN: Minnesota University Press, 2003).

—'From dead meat to glow in the dark bunnies: seeing "the animal question" in contemporary art', *parallax* 12/1 (2006), pp. 95–109.

—*What is Posthumanism?* (Minneapolis, MN: Minnesota University Press, 2009).

Wollerton, Megan, 'LG HomeChat lets you text with your appliances', *CNET*, 30 December 2003. Available at http://ces.cnet.com/8301-35306_1-57616358/lg-homechat-lets-you-text-with-your-appliances/ (accessed 23 January 2014).

Wongsrikeao, Pimprapar, Dyana Saenz, Tommy Rinkoski, Takeshige Otoi and Eric Poeschla, 'Antiviral restriction factor transgenesis in the domestic cat', *Nature Methods* 8 (2011), pp. 853–9. Available at http://www.nature.com/nmeth/journal/v8/n10/full/nmeth.1703.html (accessed 15 September 2012).

Yau, Lucy, 'Seized: Steve Kurtz and the art of dissent', *ArtVoice* 7/5 (2008). Available at http://artvoice.com/issues/v7n25/seized (accessed 19 February 2015).

Zbyszynski, Michael, 'Molecular gastronomy: spherify your food for a new culinary experience', *Make Magazine* 14 (2008), pp. 149–52.

Zurr, Ionat, 'Food for thoughts about the future', in Alex Taylor (ed.), *Experimenta House of Tomorrow: Catalogue* (Melbourne: Experimenta Media Arts, 2003).

—'The ethical claims of bioart: killing the other or self cannibalism?', *Australia and New Zealand Journal of Art: Art and Ethics* 4/2 (2003) and 5/1 (2004), pp. 167–88.

—'Growing semi-living art', PhD thesis, University of Western Australia 2009, p. 145. Available at http://theses.library.uwa.edu.au/adt-WU2009.0125/ (accessed 7 February 2014).

—and Oron Catts, 'Are the semi-living semi-good or semi-evil?' *Technoetic Arts: A Journal of Speculative Research* 1/1 (2003), pp. 47–60.

Zylinska, Joanna, *Bioethics in the Age of New Media* (Cambridge, MA: MIT Press, 2009).

Index